A Catalogue of British and Irish Prehistoric Bronzes in the Royal Ontario Museum

A Catalogue of British and Irish Prehistoric Bronzes in the Royal Ontario Museum

Francis Pryor

With drawings by Jirina Hosek and the author
and metal analyses by Dr. Paul Craddock of
the British Museum Research Laboratory

ROM

Royal Ontario Museum
Toronto

FRANCIS PRYOR is Assistant Curator in the Greek, Roman, and Prehistoric European Department.

This book has been published with the help of a grant from the Social Science Federation of Canada, using funds provided by the Social Sciences and Humanities Research Council of Canada.

Cover: Late Bronze Age shield dredged from the river Lea, front view. See catalogue number 163. Photo by L. Warren, ROM.

Canadian Cataloguing in Publication Data

Royal Ontario Museum.

A catalogue of British and Irish prehistoric bronzes in the Royal Ontario Museum

Bibliography: p.

ISBN 0-88854-240-2

1. Bronzes, British—Ontario—Toronto—Catalogs.
2. Bronzes, Irish—Ontario—Toronto—Catalogs.
3. Bronzes, Ancient—Ontario—Toronto—Catalogs.
4. Royal Ontario Museum. I. Pryor, Francis, date.

II. Title.

NK7943.R68 1980 739'.512'093610740113541 C80-094173-X

100 Queen's Park, Toronto, Canada M5S 2C6
ISBN 0-88854-240-2
Printed and bound in Canada at the ALGER PRESS

Preface

The aim of this catalogue is to bring archaeologists' attention to the most remarkable collection of British and Irish prehistoric bronzes in North America, most of which are either forgotten or have never been adequately published. The catalogue is not intended to be a visitor's guide, but has been written to give scholars information of practical use—in response to recent appeals to museums by, among others, Coles (1966) and Davey (1973:57). I have therefore decided that, with a few exceptions, I shall not attempt to fit the many artifact types described below into local typological categories. This decision was made largely because today regional classification tends to change rapidly and can only be adequately assessed if one has detailed local knowledge of the many regions and time periods represented in the collection. I have also decided to offer very full descriptions of each item's surface condition, colour, and patina, for although there is no substitute for handling individual pieces, their location here in Canada must present practical difficulties to students working in Europe. As a general policy, therefore, I have stressed description, saleroom history, and provenance reliability, in the hope that, for example, hoards that were split up in the last century may now be reconstituted, if only on paper. The aims of this catalogue are admittedly modest, but can only be achieved if the data offered are comparable with that already existing (most of it in Britain and Ireland), and I am therefore very grateful to the British Museum Research Laboratory for most generously performing the metal analyses which have added greatly to the value of this volume.

By far my greatest debt of gratitude is to Mrs. Jirina Hosek who drew most of the artifacts during the summers of 1973 and 1974. I am also most grateful to Dr. Paul Craddock, who took samples in Canada and analysed them in the British Museum Research Laboratory, and to Dr. Ian Kinnes who helped set up the analysis program. Dr. David Coombs and Dr. George Eogan read drafts of the manuscript and spent much time with me discussing various aspects of the work. Finally, my thanks are due to Ms. Sonja Tanner of the ROM Registration Department who provided me with detailed notes on the British antiquities dealers.

Francis Pryor
Toronto, October 1977

To Dr. A.D. Tushingham, F.R.S.C., F.S.A.,

formerly Chief Archaeologist,

the Royal Ontario Museum.

Contents

Introduction

THE ACQUISITION OF THE COLLECTION The prehistoric British and Irish material in the collections of the ROM was mostly acquired in the first three decades of this century by Dr. C.T. Currelly, first Director of the then University Museum of Archaeology. Currelly spent much time in London in the years before and after the Great War, and there established a number of personal contacts in the flourishing antiquities trade. This period saw the disposal of many private collections, some of which Currelly was able to purchase, in whole or in part, either direct, or, more usually, through agents.

Currelly's autobiography gives a full account of the way in which he built up the Museum's collections and from this it is apparent that he placed great emphasis on Canada's imperial ties with Britain (Currelly 1956: chap. 21). He was very keen to obtain British material of all periods. When writing to his agents in London he would first specify in which areas he thought the collections could be strengthened and would then give them a free hand to buy as they saw fit, provided they stayed within broad financial limits. He also attached great importance to quality over quantity, an attitude that was in part dictated by the shortage of storage space then available in Toronto. The success of his acquisition policy may be judged by glancing through this catalogue, where, with the exception of Early Bronze Age daggers, there is hardly a single bronze type of importance not represented.

Currelly relied heavily on the contributions of local people to pay for his purchases and most of the material described in this catalogue was obtained through the generosity of Mrs. H.D. Warren and Mr. Z.A. Lash.

At this point it is appropriate to say a few words, based largely on Museum archives, on the dealers and collectors who figure in the catalogue. The Provenance Index shows that the ROM possesses many objects dredged from the Thames. These were obtained from two sources. Mr. Llewellyn Treacher of Twyford, Berkshire, accumulated a large collection comprising palaeoliths from the Maidenhead gravels and a series of bronzes from the Thames. This collection was bought by the Museum in 1907. Treacher was a serious collector who appreciated the importance of accurate provenance for the objects in his possession—many of which he found himself.

The second source of Thames material was the antiquarian and antiquities dealer, G.F. Lawrence. Lawrence had many contacts among the contractors and workmen dredging the river and had ample opportunity to visit them when he was Inspector of Excavations for the London Museum. Like Treacher, Lawrence very much appreciated the importance of accurate provenances, which he obtained direct from the men who made the finds. The ROM bought material from Lawrence throughout the 1920s and made the last purchase of prehistoric British material, from him, in 1935.

Currelly was obviously a familiar figure to the London antiquities dealers and managed to establish an excellent rapport with most of them. One reason for this may have been that he undertook not to resell anything he bought from them on the open market. This inevitably helped to keep prices firm. London dealers from whom he obtained British bronzes included the Fenton brothers, who are discussed under the Harrogate hoard (nos. 174–176) below, and Mr. Thomas Sutton. Sutton gave the Museum his small private collection in 1922. Both these dealers were reputable and as Currelly always sought well-provenanced pieces, it may be assumed that they supplied what he required.

The largest private collection of British and Irish antiquities obtained by the ROM was that of Dr. W. Allen Sturge. His flints (including Canon Greenwell's collection) went to the British Museum, as Greenwell had requested, but everything else, including a large collection of British and Continental bronzes, was bought by the ROM in 1918, a year before Sturge's death. Sturge spent much of his life practising medicine in Nice where he was able to build up a considerable collection of classical and prehistoric European antiquities (Parkes Weber 1936). He retired to Icklingham in his native Suffolk and there resumed collecting, but concentrated in his later years on British material. His collection includes a number of Irish pieces of the Early Christian and Early Medieval periods, which have recently been published (Pryor 1976). Sturge's provenances are usually reliable, but so large a collection is bound to contain a few dubious pieces (for example, nos. 184 and 185). The Irish material Sturge obtained from the Bateman collection seems to be particularly well catalogued.

The four most important pieces in the ROM collection, the Late Bronze Age shield (no. 163) and the three items of the Harrogate hoard (nos. 174–176), were bought from private collectors, in England, who are discussed in full in the catalogue. Another small, but important and well-provenanced collection, was given to the Museum in 1943 by Lady Windle, widow of the distinguished collector and antiquarian, Sir Bertram Windle.

Finally, mention should be made of the earliest ROM purchase of prehistoric Irish material. It is designated by the accession number 906.5. No original records survive, but the relevant Museum catalogue, compiled in the 1940s from papers then in existence, clearly states that this collection was "bought from the Royal Irish Academy through L. Steele, July 1906". It consists of 26 pieces (nos. 7, 10, 13, 15, 17, 19, 21, 45–52, 65–71, 75–77, and 80) which, with one exception, are all said to have come from "near Armagh". At first glance this may seem rather improbable, but the source, if it is indeed the RIA, can hardly be considered unreliable.

In sum, it is probably fair to say that this remarkably representative collection is generally well provenanced. The condition of the pieces, too, is good, although many of the axes have had their cutting edges sharpened when still in private hands.

LAYOUT OF THE CATALOGUE The descriptive catalogue is compiled in two parts. The first part lists implements in typological categories, subdivided under two headings: Britain and Ireland. The first section closes with descriptions of the two hoards, Harrogate and Wimbledon. The second part is given over to ornaments, which are also grouped under the headings Britain and Ireland. The catalogue is followed by a chapter on the metal compositional analyses, and this is followed by the Provenance Index.

A GUIDE TO FURTHER READING This very brief guide is intended for students and visitors to the Museum who may wish to read further in British and Irish later prehistory, a subject often neglected in North American universities.

Two general prehistories of Britain (Renfrew 1974) and Ireland (Eogan and Herity 1976) have recently been published and are both essential reading, and a slightly less specialized, well-illustrated introduction to Bronze Age metalwork in England has also just appeared (Langmaid 1976). Metalworking technology is discussed, with references, by Dr. Paul Craddock (below, p. 65).

Turning to the onset of the Bronze Age, the standard work on the Beaker phenomenon is that of Clarke (1970); his ideas, however, have recently been reinterpreted in a manner now widely accepted (Lanting and van der Waals 1972). The transition from Neolithic to Bronze Age is discussed by Britton (1963) and, for Ireland, by Briggs (1976a); both papers concentrate on metalworking technology and ores. In Ireland, the distribution of three important Early Bronze Age types—daggers, halberds, and axes—is discussed in detail by Harbison (1969a; 1969b). Studies of the Early Bronze Age in Britain have traditionally been concerned with the problem of the "Wessex Culture", first defined by Piggott in 1938. References to the numerous papers that have appeared on this subject will be found in Gerloff's recent study (1975). Some workers doubt whether Early Bronze Age Wessex was ever a culture at all (for example, Coles and Taylor 1971; Briggs 1976b).

Studies of the Middle Bronze Age owe much to the papers of M.A. Smith (1959) and Butler and I.F. Smith (1956), which helped establish a sound chronology for the period. The latter half of the Bronze Age sees the emergence of distinctive regional metalworking traditions, stimulated in southern Britain by contact with the Continent (Burgess 1969a). Important regional syntheses include those of Eogan (1964) for Ireland, Burgess (1968a) for northern England, Rowlands (1976) for southern Britain, and Savory (1958) for Wales. Bronze Age metalwork in Scotland is comprehensively reviewed in three papers by Coles (1959–60; 1963–64; 1968–69).

Surveys of smaller geographical areas reveal the complexity underlying the regional syntheses. Fox's pioneering detailed survey of the Cambridge region (1923) places its Bronze Age metalwork in its environmental and chronological setting.

Davey's recent (1973) study of Bronze Age metalwork in Lincolnshire is also outstanding. In the last two decades provincial museums have published catalogues that well illustrate local metalworking traditions (for example, Annable and Simpson 1964; Langmaid 1966; Moore and Rowlands 1972).

The latest Bronze Age and the whole of the Iron Age of Britain is comprehensively surveyed by Cunliffe (1974) and a similarly detailed volume, covering the same period, but for southern Britain only, has been published by Harding (1974). Both books have full bibliographies.

Even a brief survey should mention reviews devoted to life (Simpson 1971; Burgess and Miket 1976) and death (Ashbee 1960; 1970) in later prehistoric Britain. Finally, journals that usually contain important papers on British and Irish prehistory include *The Antiquaries Journal*, *Antiquity*, *Archaeologia*, *Archaeologia Cambrensis* (Wales), *The Archaeological Journal*, *Current Archaeology* (more popular), *The Journal of the Royal Society of Antiquaries of Ireland*, *The Proceedings of the Prehistoric Society*, *The Proceedings of the Royal Irish Academy*, *The Proceedings of the Society of Antiquaries of Scotland*, and *The Ulster Journal of Archaeology*.

The Catalogue

NOTES The catalogue describes every British and Irish prehistoric bronze in the Museum's collections. For the purposes of this work each item has been assigned a number which is used in both the text and the illustrations. Swords (nos. 142–153), the Harrogate hoard (nos. 174–176), and ornaments (nos. 177–184) are drawn at one-third scale and all other drawings are half the actual size. The selection of these large scales has made it necessary to depart from recognized chronological or typological series in the arrangement of the illustrations.

The ROM accession number appears at the end of each catalogue entry. These numbers contain information useful in tracing a lost object and so are briefly explained here by an example: the number **918.33.108**. It is composed of three parts. The first part, the **year number**, indicates that the item was obtained in 1918. The second part, the **acquisition number**, shows that the example was acquisition 33 of 1918. Finally, the presence of a **registration number** indicates that acquisition 33 contained more than one item, each one of which was allotted a specific number. The registration number is usually taken from invoices or sales lists and is generally not allocated afresh by the Museum. The letter "x" in the accession number indicates that the Museum no longer possesses the original papers relating to the object.

In the descriptions that follow, maximum lengths (L.) are given in centimetres and weights (W.) are given in grams.

Implements

Flat and Flanged Axes

BRITAIN

(Britton 1963; Burgess and Cowen 1972; Coghlan and Case 1957; Megaw and Hardy 1938)

1. Broad-butted flat axe. Both surfaces very rough, pitted; mottled brown/green patina mostly lifted.
L. 13.15, W. 504.4
Cornwall. Sturge colln.
918.33.108

2. Thin-butted flat axe of Migdale-Marnoch tradition. Surface slightly pitted; cutting edge severely flattened at centre; brown patina.
L. 14.75, W. 533.2
Thames. Bought from L. Treacher, Twyford.
908.32.7

3. Broad-butted flat axe. Surface slightly pitted and scratched (? in antiquity). Cutting edge blunted (? modern). Dark brown/green mottled patina.
L. 15.9, W. 890.1
Presumably English. Ex Windle colln. Windle obtained it from a Mr. Cotton of Bromsgrove (no. XLII in his colln.).
943.41.5

4. Broad-butted flat axe. Sides damaged in modern times; surface very rough; light green/dark brown mottled patina.
L. 12.8, W. 377.1
One of hoard (?) dredged from Thames, near Reading. Bought from L. Treacher, Twyford.
908.32.8

5. Flanged axe of Arreton tradition. Upper blade decoration: alternate reserved and filled chevrons; other side: chequerboard pattern of reserved and cross-hatched lozenges. Decoration on both sides bounded by zone, parallel to the cutting edge, of row of dots above three scratched lines, above low curved ridge. Both sides fluted. Very smooth brown patina lifting (halted by conservation).
L. 11.0, W. 332.5
Yorkshire. Sturge colln. Strong resemblance to Irish Derryniggin type (Harbison 1969a).
918.33.118

6. Thin-butted flat axe of Migdale-Marnoch tradition. Condition good, but surface very pitted.

Cutting edge blunt; butt damaged (? modern). Brown patina.
L. 10.55, W. 163.5
Dredged from the Thames. Bought from L. Treacher, Twyford.
908.32.9

IRELAND

(Harbison 1969a; Coghlan and Case 1957)

The classification used here is that of Harbison (1969a). The ROM possesses examples of all four major types, none of which is listed in his group whose "present whereabouts are unknown" (Harbison 1969a: 92).

Lough Ravel thick-butted flat axe (Coghlan and Case 1957), nos. 12 and 20.
Killaha flat; thin butt, broadly splayed sides (Coghlan and Case 1957), nos. 10, 21, and 22.
Ballyvalley flat/flanged; thin butt, parallel sides on upper half (Megaw and Hardy 1938: type 1 and type 2, but with lower flanges), nos. 7–9, 11, 14–19, and 23.
Derryniggin flanged; thin butt, parallel sides on upper ⅔ or ¾ (Megaw and Hardy 1938: type 2, high flanged, and type 3), no. 13.

7. Axe of Ballyvalley type with very low beaten-up flanges. Pitted surface; bronze disease halted. Cutting edge sharpened (? modern). Pale green/light brown.
L. 11.0, W. 234.1
Near Armagh. Bought through L. Steele (see Introduction).
906.5.6

8. Axe of Ballyvalley type with beaten-up flanges and low stop-ridge. Good condition, but slightly pitted. Butt and cutting edge damaged; brown patina.
L. 8.8, W. 145.3
Ballymena, Co. Antrim. Sturge colln.
918.33.112

9. Small flanged axe of probable Ballyvalley type. Surface good, but slightly uneven and cutting edge very blunt. Brown/green mottled patina.
L. 6.4, W. 77.3
Co. Down. Sturge colln.
918.33.110

10. Flat axe of Killaha type. Surface very pitted and scratched. Modern cleaning and possible sharpening of blade; brown patina.
Near Armagh. Bought through L. Steele (see Introduction).
906.5.4

11. Flanged axe of Ballyvalley type. Low cast flanges and slight stop-ridge. Very slight, punched diagonal lines on upper blade. Butt damaged. Surface and condition good, but cutting edge sharpened recently. Light brown patina. "R.J. Bark" or "R.J. Bank" written in 19th-century hand near butt.
L. 12.6, W. 340.1
Ireland. Sturge colln.
918.33.113

12. Flat axe of Lough Ravel type. Butt and blade nicked (? modern); surface rough; dark brown/green mottled patina.
L. 7.7, W. 209.7
Ireland. Sturge colln.
918.33.103

13. Flanged axe of Derryniggin type. Hammered cable decoration along both sides. Patina intact on illustrated side; other side pitted; traces of recent sharpening; dark brown patina.
L. 8.0, W. 86.6
Near Armagh. Bought through L. Steele (see Introduction).
906.5.8

14. Flat axe of Ballyvalley type. Butt hammered flat. Possible signs of hafting (transverse striations on upper haft and deep horizontal nicks on both sides); cutting edge very blunt. Few deep pits, otherwise surface smooth, pale brown.
L. 9.35, W. 171.7
Co. Down. Sturge colln.
918.33.108

15. Flat axe of Ballyvalley type. Condition good, but blade shows (? modern) abrasion; cutting edge very blunt; brown/green mottled patina.
L. 10.5, W. 119.4

Near Armagh. Bought through L. Steele (see Introduction).
906.5.7

16. Flat axe of Ballyvalley type. Hammered (? punched) herringbone decoration along blade sides on both faces; sides have light horizontal decoration. Condition excellent; patina brown/green mottled.
L. 14.4, W. 377.3
Armoy, Co. Antrim. Sturge colln.
918.33.104

17. Flanged axe of Ballyvalley type. Flanges cast. Blade and cutting edge sharpened (? modern); butt damaged (? modern). Dark brown patina; condition fair.
L. 16.0, W. 463.5
Near Armagh. Bought through L. Steele (see Introduction).
906.5.2

18. Flanged axe of Ballyvalley type. Light stop-ridge; cast flanges. Condition good, surface lightly pitted; patina brown/dark brown.
L. 15.05, W. 397.3
Ireland. Sturge colln.
918.33.111

19. Flanged axe of Ballyvalley type. Very low stop-ridge and low beaten-up flanges. Extremely faint "herringbone fluting" on blade; similar style vertical lines above stop-ridge. Sharp cutting edge. Surface condition smooth; sides pitted; brown patina.
L. 12.9, W. 278.3
Near Armagh. Bought through L. Steele (see Introduction).
906.5.5

20. Flat axe of Lough Ravel type. Cutting edge nicked and butt damaged (? modern). Surface rough; patina brown/green mottled.
L. 11.3, W. 491.0
Sligo. Sturge colln.
918.33.108

21. Flat axe of Killaha type. Condition excellent; cutting edge and butt slightly worn; patina dark brown.
L. 13.0, W. 348.0
Near Armagh. Bought through L. Steele (see Introduction).
906.5.3

22. Flat axe of Killaha type. Condition good. Surface slightly pitted, with vertical (? modern) striations and very faint horizontal hammered ripples near butt.
L. 14.25, W. 566.3
NW Ireland. Sturge colln.
918.33.106

23. Flanged axe of Ballyvalley type. Low flanges and stop-ridge. Modern damage to cutting edge; apparent "cable" pattern is modern damage. Brown/light brown patina.
L. 11.4, W. 238.6
Ireland. Sturge colln.
918.33.114

Palstaves and Winged Axes

(Classification based on Smith 1959)

Britain

24. Looped wing-flanged axe. Rounded shield pattern. Stop-ridge lower on one face than on other. Ancient damage: butt, one wing, and left side of blade. Modern damage: horizontal cuts high on both sides. Smooth olive green patina survives in patches.
L. 12.8, W. 425.4
No provenance. Bought from T. Sutton, 1927.
927.27.27

25. Unlooped low-flanged palstave. Flared cutting edge and median rib below stop-ridge. Deep casting flaw just above stop-ridge on side now shown. Cutting edge and blade sharpened recently; pale green patina scored by crude cleaning. Legend "A.I." and "Beck" (or "Bech") in 19th-century hand near butt.
L. 16.20, W. 436.0
Sussex. Sturge colln.
918.33.122

26. Looped palstave of late type. Smooth, dark patina, greener and rougher towards cutting edge. Condition excellent.

L. 13.7, W. 287.7
Near York. Bought from G.F. Lawrence, Oct. 6, 1928.
928.35.7

27. Unlooped palstave of Portrieux type. Wide median rib below stop-ridge. Left corner of blade broken and butt damaged in antiquity. Surface rough, pitted on blade; dark/pale green patina.
L. 17.1, W. 467.1
Cornwall. Sturge colln. Find-spot consistent with the southerly British distribution of this Breton type (Burgess 1970).
918.33.129

28. Unlooped low-flanged palstave. Shield pattern with pendant median ridge. Jagged cutting edge. Faint horizontal lines on sides. Surface pitted; light green patina.
L. 14.6, W. 339.7
Norfolk. From W.S. Simpson (?).
908x34.2

29. Looped low-flanged palstave. Median rib below stop-ridge. Butt heavily hammered; corners and blade of cutting edge battered. Patina lost.
L. 12.35, W. 384.4
Thames. From L. Treacher, Twyford.
908.32.5

30. Unlooped wing-flanged axe. Cutting edge damaged in antiquity; one wing cracked; butt hammered. Dark surface, lightly pitted.
L. 12.6, W. 321.2
Thames. Bought from L. Treacher, Twyford.
908.34.4

31. Unlooped low-flanged palstave. Shield pattern with median rib. Letter "B" and "Ely" in white ink on butt. Cutting edge sharp, undamaged. Brown/green mottled patina, rough surface.
Ely, Cambridgeshire. Sturge colln.
918.33.128

32. Unlooped haft-flanged axe. Flange nicked near stop-ridge; uneven butt; slightly blunted cutting edge. Surface smooth; light brown patina almost intact.
L. 16.0, W. 482.6
Lakenheath Fen, Suffolk. No data on purchase.
908x34.1

33. Small, unlooped low-flanged palstave. Slight, wide median rib below stop-ridge. Surface light green, sandy feel (? patina lost).
L. 7.8, W. 130.9
Cornwall, Sturge colln.
918.33.124

34. Unlooped palstave of transitional type. Stop-ridge interrupted at flanges to form central square projection. Very dark brown patina over uneven surface. Butt damaged. Iron-pan (from sand or gravel subsoil ?) adheres to surface.
L. 9.6, W. 189.4
Essex. Sturge colln.
918.33.120

35. Looped palstave of southwestern type. Deeply grooved V-shaped shield pattern on both faces. Butt and blade damaged. Dark green patina intact.
L. 15.7, W. 479.4
Sussex. Gift of Mr. Thomas Sutton, Dec. 18, 1922.
922.37.1

36. Unlooped low-flanged palstave. Wide, low median rib below stop-ridge. Patina lost, surface pitted.
L. 14.85, W. 406.7
Thames at Maidenhead. From L. Treacher, Twyford.
908.32.6

37. Unlooped low-flanged palstave. Shield pattern with median rib. Cutting edge recently sharpened, otherwise condition excellent; patina black.
L. 15.95, W. 448.6
Thames. Bought from S.G. Fenton.
927.51.24

38. Unlooped low-flanged palstave. Slight, flat, median rib below stop-ridge. Corner of butt damaged; cutting edge recently sharpened. Dark brown patina with (?) iron staining above stop-ridge.
L. 17.1, W. 418.8
Mitcham, Surrey. Purchased from G.F. Lawrence, Jul. 15, 1929.
929.16.15

Ireland

39. Unlooped low-flanged palstave of late type. Pale green patina intact on butt only.
L. 12.7, W. 360.8
Co. Antrim. Sturge colln.
918.33.123

40. Haft-flanged axe. Very thin stop-ridge and short, low median rib. Butt hammered-over. Surface pitted; dark mottled green patina.
L. 14.15, W. 338.0
Ireland. Sturge colln.
918.33.119

41. Looped palstave of late type. Butt damaged (? recently). Cleaned and conserved; surface slightly rough; cutting edge sharp.
L. 10.7, W. 193.5
Armoy, Co. Antrim. Sturge colln.
918.33.131

42. Unlooped palstave (?) with very low flanges and stop-ridge. Most unusual form. Clear scratches on sides where excess metal has been removed. Condition good; surface smooth; brown patina.
L. 11.35, W. 163.0
Ireland. Sturge colln. *Modern* (see analyses below).
918.35.115

43. Small unlooped palstave. Cutting edge recently sharpened. Surface smooth, but flange sides rough; patina dark brown.
L. 7.05, W. 118.0
Co. West Meath, 1856. Ex Brayton colln.
913x11.27

44. Damaged unlooped palstave. Blade and half butt only. Damaged in antiquity. Cutting edge sharpened recently. Surface smooth; patina dark reddish brown.
L. 8.2, W. 196.7
Co. Down. Sturge colln.
918.33.125

45. Wing-flanged axe. Shield pattern with median rib. Cutting edge sharpened recently. Dark reddish brown patina largely lost.
L. 13.1, W. 389.0
Near Armagh. Bought through L. Steele (see Introduction).
906.5.24

46. Haft-flanged axe. Short flange; no stop-ridge. Butt heavily hammered; blunt cutting edge; surface good; patina dark brown.
L. 13.3, W. 321.2
Near Armagh. Bought through L. Steele (see Introduction).
906.5.22

47. Haft-flanged axe. Slight trace of vertical lines on surviving patina below stop-ridge. Cutting edge recently sharpened; sides damaged recently. Surface pitted; butt scored (ancient); brown patina largely lost.
L. 15.3, W. 511.5
Near Armagh. Bought through L. Steele (see Introduction).
906.5.20

48. Haft-flanged axe. Thin stop-ridge; flanges slightly damaged. Cleaned, but traces of dark brown patina survive; surface smooth, with few slight pits.
L. 13.35, W. 358.6
Near Armagh. Bought through L. Steele (see Introduction).
906.5.23

49. Wing-flanged axe. Low relief crescentic shield pattern. Butt hammered. Blade recently sharpened. Surface smooth, except for deep pits on upper blade; sides less smooth than blade. Dark brown patina.
L. 15.2, W. 676.7
Near Armagh. Bought through L. Steele (see Introduction).
906.5.21

50. Wing-flanged axe. Butt damaged and wings nicked in antiquity. Cutting edge recently sharpened. Dark brown patina intact, but sides more corroded.
L. 15.95, W. 565.2
Near Armagh. Bought through L. Steele (see Introduction).
906.5.19

51. Looped wing-flanged axe. Long, high boss in

centre of stop-ridge; wings hammered-over. Butt truncated and blade deeply grooved on both sides in antiquity; otherwise condition good; dark brown patina.
L. 11.7, W. 362.3
Near Armagh. Bought through L. Steele (see Introduction).
906.5.25

52. Small, looped haft-flanged axe. Deep stop-ridge cut by two V-shaped grooves and face of protruding central boss decorated with light diagonal grooves. Slight trace of (?) decorative grooves on blade, near cutting edge; horizontal tooled grooves on sides near butt. Side loop broken. Condition excellent; surface smooth; patina dark brown.
L. 8.7, W. 221.6
Near Armagh. Bought through L. Steele (see Introduction). *Modern* (see analyses below).
906.5.26

Socketed Axes

BRITAIN

(Burgess 1969a)

53. Looped socketed axe of Breton type. Blunt cutting edge. Surface smooth; grey green patina.
L. 12.55, W. 308.1
Probably English. Ex colln. of Sir Bertram Windle; given to ROM by Lady Windle, Aug. 1943.
943.41.2

54. Looped socketed axe of Sompting type (Burgess 1969b). Cutting edge damaged in antiquity; rim top damaged (modern). Surface rough, pitted; patina dark brown.
L. 10.85, W. 242.4
Near Bradford, Yorkshire. Bought from G.F. Lawrence, Dec. 5, 1929.
929.16.3

55. Looped socketed axe. Cutting edge blunted. Surface smooth, but with patches of pits and encrustation; patina dark green.
L. 7.05, W. 135.4
Kingston Buci, Sussex. Bought from G.F. Lawrence, Nov. 9, 1929.
929.16.18

56. Damaged looped socketed axe of Yorkshire type. Three widely spaced vertical ribs, typical of the type, pendant from horizontal rib below collar. Lower blade broken off in antiquity. Patina, olive green and smooth, survives near rim.
L. 6.7, W. 142.1
North Wales. Sturge colln.
918.33.136

57. Looped socketed axe of Yorkshire type. Two clear and one very faint central rib pendant from horizontal rib below collar; three widely spaced ribs on other side. Casting flaw below collar. Cutting edge unused. Condition excellent; patina dark green, mottled.
L. 7.9, W. 170.8
Near Boston, Lincolnshire.
908x34.5

58. Looped socketed axe of Breton type. Very pronounced external casting ridges; socket rim not trimmed. Cutting edge unused. Surface pitted; patina mottled green.
L. 13.35, W. 238.5
Sussex.
908x34.7

59. Looped socketed axe of Sompting type (Burgess 1969b). Flared, crescentic cutting edge; three pendant ribs from horizontal rib below collar. Cutting edge sharpened recently. Surface smooth; patina dark green.
L. 12.0, W. 353.3
Yorkshire. Sturge colln.
918.33.137

60. Bag-shaped looped socketed axe. Low collar; socket damaged, probably the result of use; cutting edge blunted. Lightly pitted surface; patina dark brown.
L. 8.85, W. 213.6
Derbyshire. Sturge colln.
918.33.133

61. Looped socketed axe. Four short ribs on socket internal walls. Sharp cutting edge. Surface smooth; pale brown patina.
L. 9.95, W. 199.7
Thames near Maidenhead.
Bought from L. Treacher, Twyford.
907.16.3

62. Looped socketed axe of Continental type. Decoration: three horizontal ribs below collar; blade decorated with ribbed Y-shaped motif. Middle and lower blade damaged (? in antiquity). Very dark brown/black patina.
L. 11.5, W. 257.2
Tillingham, Essex. Sturge colln.
918.33.135

63. Looped socketed axe. Sharp cutting edge; metal cracked around loop and mouth (? from use). Surface smooth; patina very dark brown.
L. 9.8, W. 205.0
Near Boston, Lincolnshire.
908x34.4

64. Looped socketed axe of Yorkshire type. Three widely spaced vertical ribs pendant from horizontal rib below collar. Sharp cutting edge. Surface smooth; patina brown/light brown mottled.
L. 8.35, W. 200.3
"Welwyn, Leamington, England." [sic].
908x34.3

IRELAND

(Eogan 1964)

65. Looped bag-shaped socketed axe. Cutting edge slightly worn. Condition excellent; surface smooth; patina dark brown.
L. 7.6, W. 144.3
Near Armagh. Bought through L. Steele (see Introduction).
906.5.11

66. Looped bag-shaped socketed axe. Worn cutting edge. Condition excellent; smooth surface; dark green patina.
L. 5.6, W. 124.6
Near Armagh. Bought through L. Steele (see Introduction).
906.5.13

67. Looped bag-shaped socketed axe. Condition excellent, but cutting edge recently sharpened. Surface quite rough; patina brown.
L. 10.35, W. 438.1
Near Armagh. Bought through L. Steele (see Introduction).
906.5.9

68. Looped bag-shaped socketed axe with enlarged collar. Two narrow horizontal ribs below collar flank wide raised band. Right side of blade damaged, otherwise condition good; surface slightly rough; patina dark brown; sharp cutting edge.
L. 7.6, W. 165.6
Near Armagh. Bought through L. Steele (see Introduction).
906.5.12

69. Looped bag-shaped socketed axe. Decoration as no. 68 above. Cutting edge recently sharpened. Surface uneven; condition fair; patina green.
L. 9.7, W. 275.3
Near Armagh. Bought through L. Steele (see Introduction).
906.5.1

70. Looped bag-shaped socketed axe. Flared mouth. Casting ridge flared on loop side, smoothed on other side. Cutting edge worn, irregular. Surface texture uneven; smooth dark patina with areas of pitting caused by uneven mould surface.
L. 9.3, W. 327.8
Co. Down. Probably bought through L. Steele (see Introduction).
906.5.10

71. Small looped socketed axe. Thin transverse section. Thick collar. Sharp cutting edge. Surface smooth, pitted in places; mid-brown patina.
L. 5.6, W. 83.4
Near Armagh. Bought through L. Steele (see Introduction).
906.5.17

72. Looped bag-shaped socketed axe. Top of rim and external casting seams filed flat with a rough rasp. Cutting edge battered. Condition excellent; slightly rough surface; very dark brown patina. Bought in a second-hand shop in Cork and almost certainly from Co. Cork. Ex colln. of Sir Bertram Windle; given to ROM by Lady Windle, Aug. 1943. *Modern* (see analyses below).
943.41.2

73. Looped bag-shaped socketed axe. Cutting edge blunted in antiquity; metal cracked opposite loop. Surface slightly pitted; dark green patina.

L. 7.7, W. 199.7
Cork. Sturge colln.
918.33.132

74. Unlooped bag-shaped socketed axe. Pitted surface; patina lost. Left corner of blade battered.
L. 6.5, W. 122.0
Bought in a second-hand shop in Cork and almost certainly from Co. Cork. Ex colln. of Sir Bertram Windle; given to ROM by Lady Windle, Aug. 1943.
943.41.2

75. Looped bag-shaped socketed axe. Two very faint horizontal grooves below rim. Cutting edge recently sharpened. Surface rough; mottled brown patina.
L. 5.75, W. 88.5
Near Armagh. Bought through L. Steele (see Introduction).
906.5.16

76. Looped bag-shaped socketed axe. High, sharp internal V-shaped transverse ridge starts 15 mm below rim. Sharp cutting edge. Surface very smooth; patina dark brown/black.
L. 6.2, W. 91.0
Near Armagh. Bought through L. Steele (see Introduction).
906.5.15

77. Looped socketed axe. Two slight horizontal ribs below collar. Cutting edge sharp. Patina, dark green, intact around socket and loop.
L. 6.7, W. 138.8
Near Armagh. Bought through L. Steele (see Introduction).
906.5.14

78. Looped bag-shaped socketed axe. Cutting edge blunt; red brown patina largely lost.
L. 7.8, W. 169.5
Antrim. Sturge colln.
918.33.134

79. Looped socketed axe. Three widely spaced ribs pendant from horizontal rib below collar. Flared cutting edge, sharpened recently. Surface rough, not pitted; patina light green/brown mottled.
L. 8.1, W. 213.9
Ireland. Sturge colln.
918.5.44

80. Small looped bag-shaped socketed axe. Cutting edge recently sharpened. Traces of original brown patina.
L. 5.3, W. 48.3
Near Armagh. Bought through L. Steele (see Introduction).
906.5.18

Spearheads

BRITAIN

(Burgess 1969a; Greenwell and Brewis 1909; Smith 1959)

81. Ribbed kite-shaped spearhead with socket loops and peg-holes. One loop damaged. Base of blade much damaged giving false leaf-shaped outline. Surface pitted, but dark brown patina survives along central ridge.
L. 15.3, W. 104.2
Thames. Sturge colln.
918.33.1

82. Leaf-shaped spearhead with socket loops. Broken below blade (restored). Poor condition; green/brown mottled patina.
L. 15.9, W. 84.5
Maidenhead (encrustation suggests the Thames). Bought from L. Treacher, Twyford.
907.16.4

83. Leaf-shaped spearhead with socket loops. Flattened socketed loops (cf. no. 82 above). Very dark brown patina largely lost.
L. 12.4, W. 77.3
Thames at Sion Reach. Bought from G.F. Lawrence, Jul. 15, 1929.
929.16.14

84. Leaf-shaped spearhead with socket loops. Socket damaged on side not drawn. Dark green smooth patina; surface pitted more heavily on side not illustrated.
L. 10.3, W. 49.8
Suffolk.
908x34.9

85. Leaf-shaped spearhead with peg-holes.

Smooth surface where dark brown patina survives. Socket extends to tip.
L. 22.2, W. 172.5
Thames at Windsor. Bought from G.F. Lawrence, Mar. 1, 1927.
927.28.100

86. Leaf-shaped spearhead with peg-holes. Socket extends to tip. Blade bent. Peg-holes filled with iron encrustation. About 1/4 of both surfaces covered by pale calcareous encrustation. Patina black, where present.
L. 28.1, W. 238.8
Thames at Windsor. Bought from G.F. Lawrence, Feb. 15, 1927.
927.28.99

87. Leaf-shaped spearhead with peg-holes. Socket squashed below blade. Socket extends to tip. Surface smooth, dark brown patina; blade edges slightly damaged.
L. 17.3, W. 110.3
Thames near Maidenhead. Bought from L. Treacher, Twyford.
907.16.8

88. Leaf-shaped spearhead with peg-holes. Socket extends to tip. Black/dark brown patina mostly intact; surface smooth.
L. 16.3, W. 104.7
Thames at Reading. Bought from W. Talbot-Ready, Sept. 17, 1907.
907.14.4

89. Leaf-shaped spearhead with peg-holes. Socket extends through to tip. Casting flaw near socket mouth. Brown slightly pitted surface with much coarse sandy encrustation.
L. 14.1, W. 81.1
Thames near Reading. From L. Treacher, Twyford. Note: this spearhead and nos. 90, 91, and 94 are said to have belonged to a hoard dredged from the Thames. This association is rather dubious, especially bearing in mind the extremely large numbers of bronzes that were dredged from the river in the years before and after World War I (Lawrence 1929). The surface condition of the four pieces tends to support the view that the hoard is more apparent than real.

90. Leaf-shaped spearhead with peg-holes. Socket extends through to tip. Whitish green/brown patina with sandy encrustation (not illustrated), largely removed, still covering 2/3 of surface.
L. 16.5, W. 114.0
Thames near Reading. From L. Treacher, Twyford. Hoard? See note to no. 89 above.
907.16.9

91. Blade of leaf-shaped socketed spearhead. Half of one surface covered with chalky encrustation (not illustrated); other surface minutely pitted, with sandy encrustation.
Thames near Reading. From L. Treacher, Twyford. Hoard? See note to no. 89 above.
907.16.11

92. Leaf-shaped spearhead with peg-holes. Socket extends through to tip. Whitish green patina with coarse sandy encrustation on 1/5 of both surfaces (not illustrated). Condition excellent, surface smooth.
L. 12.6, W. 78.7
Thames near Reading. Bought from L. Treacher, Twyford.
907.16.6

93. Leaf-shaped spearhead with peg-holes. Socket extends through to tip. Whitish green/brown patina with coarse sandy encrustation over ca 1/4 of both surfaces (not illustrated). Condition excellent; surface smooth.
L. 11.2, W. 64.7
Thames at Reading. Bought from L. Treacher, Twyford.
907.16.7

94. Leaf-shaped spearhead with peg-holes. Socket extends through to tip. Socket damaged near tip on side not drawn. Surface smooth, but whitish green patina lightly covered with even, thin spread of sandy encrustation (not illustrated).
L. 11.9, W. 53.7
Thames near Reading. Bought from L. Treacher, Twyford. Hoard? See note to no. 89 above.
907.16.12

95. Leaf-shaped spearhead with peg-holes. Socket extends through to tip. Smooth surface with whitish green patina. About 4/5 of both surfaces covered with sand, gravel, and shell concretion (not illustrated). Encrustation possibly heavier on one surface.

L. 9.7, W. 69.1
Thames near Reading. Bought from L. Treacher, Twyford.
907.16.15

96. Leaf-shaped spearhead with peg-holes. Socket extends through to tip. Whitish green/brown patina with thin scatter of coarse sand concretion over ca $^{1}/_{10}$ of both surfaces (not illustrated). Surface smooth.
L. 10.0, W. 61.2
Thames near Reading. Bought from L. Treacher, Twyford.
907.16.14

97. Damaged leaf-shaped socketed spearhead with low ribs. Socket extends through to tip. Whole of socket and parts of lower blade damaged by intense heat. The small holes in the drawing are probably *not* peg-holes. Whitish green patina with coarse sand and shell concretion over ca $^{1}/_{5}$ of surface (not illustrated). Surface smooth.
L. 13.9, W. 78.6
Thames near Reading. Bought from L. Treacher, Twyford.
907.16.9

98. Fragment of socketed spearhead. Brown patina; surface smooth.
W. 3.2
Thames near Reading. Bought from L. Treacher, Twyford.
907.16.34

99. Fragment of socketed spearhead. Surface smooth. Brown patina with slight calcareous encrustation.
W. 2.5
Thames near Reading. Bought from L. Treacher, Twyford.
907.16.35

100. Leaf-shaped spearhead with peg-holes. Socket extends through to tip. Extensive damage to socket near tip on side not drawn. Whitish green/brown patina; smooth surface; blade edges battered.
L. 11.8, W. 56.6
Thames near Reading. Bought from L. Treacher, Twyford.
907.16.3

101. Damaged socketed spearhead with peg-holes. Probably leaf-shaped. Lower blade and upper socket damaged by intense heat. Most of blade missing. Undamaged surfaces smooth; whitish green-brown patina covered over ca $^{1}/_{5}$ of surface by sandy concretion (not illustrated).
L. 7.9, W. 58.6
Thames near Reading. Bought from L. Treacher, Twyford.
907.16.22

102. Damaged leaf-shaped socketed spearhead with one peg-hole surviving. Socket extends through to tip. Socket damaged near tip and severely distorted and damaged by intense heat near mouth. Inside of socket filled with sand concretion. Surface smooth where undamaged; pale brown patina.
L. 9.5, W. 66.0
Thames near Reading. Bought from L. Treacher, Twyford.
907.16.17

103. Damaged leaf-shaped socketed spearhead with one peg-hole surviving. Socket extends through to tip. Lower part of blade and socket buckled and damaged by heat. Whitish green/brown patina with coarse sand encrustation over ca $^{1}/_{3}$ of surface (not illustrated).
L. 10.8, W. 58.3
Thames near Reading. Bought from L. Treacher, Twyford.
907.16.20

104. Fragment of socketed spearhead with peg-holes. Probably leaf-shaped, with socket extending through to tip. Buckled and damaged by intense heat. Whitish green patina; uneven heat-affected surface with coarse sandy concretion over ca $^{2}/_{5}$ of surface (not illustrated).
L. 10.5, W. 79.2
Thames near Reading. Bought from L. Treacher, Twyford.
907.16.21

105. Damaged leaf-shaped socketed spearhead with peg-holes. Tip cleanly broken and socket squashed, to give oval section. Whitish green patina with very slight sandy concretion; colour browner on one side.
L. 8.3, W. 48.3

Thames near Reading. Bought from L. Treacher, Twyford.
907.16.16

106. Damaged socketed spearhead. Probably leaf-shaped. Socket extends through to tip. Socket lost below blade. Blade severely buckled by heat. Whitish green patina with much (ca 2/3 of surface) coarse sand and gravel concretion (not illustrated).
L. 9.3, W. 77.5
Thames near Reading. Bought from L. Treacher, Twyford.
907.16.23

107. Blade of leaf-shaped socketed spearhead. Hollow socket extends 6 mm into blade. Broken in antiquity. Dark green patina; surface smooth and lustrous.
L. 9.1, W. 58.6
From a barrow, Unchfont Down, Wiltshire, 1893.
908x32.10

108. Damaged leaf-shaped spearhead with peg-holes. Most of blade cleanly broken off in antiquity. No signs of heat. Whitish green/brown patina with fine sandy concretion over ca 1/5 of surface (not illustrated); surface otherwise smooth.
L. 6.8, W. 56.2
Thames near Reading. Bought from L. Treacher, Twyford.
907.16.18

109. Leaf-shaped socketed spearhead. Socket extends through to tip. Excellent condition; smooth surface; dark green patina.
L. 15.9, W. 89.3
Probably from Suffolk. From G.F. Lawrence.
909.67.2

110. Leaf-shaped spearhead with peg-holes. Socket extends through to tip. Socket below blade seriously damaged. Surface damaged in antiquity. Edges of blade battered. Smooth, brown/dark brown patina. Label in 19th-century hand, "Bronze spearhead found at the foot of the Wrekin, Salop. Given to one by Pe. Thursfield Esq. Broseley."
L. 18.8, W. 136.8
The Wrekin, Shropshire. Ex colln. of Sir Bertram Windle; given to ROM by Lady Windle, Aug. 1943.
943.41.6

111. Subtriangular basal-looped socketed spearhead. Socket extends through to tip. Patina very dark brown/black; surface smooth. Side not drawn has fine sandy encrustation over 1/5 of its surface. Condition excellent.
L. 37.0, W. 261.7
The Thames at Battersea, 1858. Sturge colln. (ex Bateman colln.). Mentioned in Evans (1881:327).
918.33.2

112. Subtriangular basal-looped socketed spearhead. Socket extends through to tip. Blade broken and edges battered, otherwise condition good. Surfaces smooth with patina dark brown on one side, lighter brown on other.
L. 33.1, W. 261.1
Mildenhall, Suffolk. Bought by G.F. Lawrence at Fenton Sale no. 13, Dec. 30, 1927.
927.3.5

113. Leaf-shaped basal-looped socketed spearhead. Two low ribs parallel to median rib. Very dark brown patina, largely intact.
L. 24.6, W. 208.9
Thames at Hammersmith. Bought from G.F. Lawrence, Apr. 9, 1929.
929.16.11

Ireland

(Eogan 1964)

114. Ribbed kite-shaped spearhead with socket loops. One loop damaged; socket damaged on side not drawn. Loop wide, flattened. Surface rough; patina dark brown/black.
L. 21.0, W. 144.0
Ireland. Sturge colln.
918.33.100

115. Ribbed kite-shaped spearhead with socket loops. Socket extends halfway along blade. Cleaned, surface smooth; condition excellent.
L. 9.9, W. 64.4
Ireland.
913x11.14

116. Kite-shaped spearhead with socket loops at base of blade. Socket extends through to tip. Long sides of blade heavily nicked. Brown/dark brown patina; surface smooth, with a few shallow pits.

L. 15.3, W. 70.4
Ireland. Sturge colln.
918.33.101

117. Leaf-shaped spearhead with socket loops. Socket extends well into blade, but not to tip. Blade small, very thin. Socket mouth extensively nicked (? modern).
L. 12.4, W. 49.2
Rathkeale, Co. Limerick.
908x34.3

118. Kite-shaped spearhead with socket loops. Socket extends through to tip. Slight damage to left-hand corner of blade and loop below. Patina dark brown and surface smooth.
L. 9.0, W. 46.0
Ireland.
913x11.16

119. Spearhead with socket loops and very reduced blade. Socket continues halfway to tip. Very thin socket loops and irregular blade outline suggest that this is a mis-casting. Surface smooth; dark brown patina.
L. 10.9, W. 37.8
Ireland.
913x11.15

120. Leaf-shaped socketed spearhead. Single peg- or pinhole made after casting. Socket extends through to tip. Surface smooth; patina dark brown.
L. 9.9, W. 38.5
Ireland.
913x11.13

121. Leaf-shaped socketed spearhead with two lunate openings in blade. Two off-centre pin or peg-holes near socket mouth made after casting with square-sectioned punch. Socket extends ca halfway to tip. Surface smooth; patina dark brown.
L. 14.9, W. 57.1
Ireland.
913x11.12

122. Leaf-shaped, hollow, ribbed spearhead with two peg-holes. Tip broken to reveal hollow interior. Socket rim much damaged and possibly truncated. Surface smooth, with very small pits; patina brown, lustrous.
L. 17.8, W. 185.7
Ireland.
913x11.10

123. Leaf-shaped spearhead with peg-holes. Socket extends to tip. Socket mouth bent inward and damaged, possibly truncated. Edge of blade nicked; surfaces cleaned, evenly pitted.
L. 29.4, W. 143.2
Ireland.
913x11.8

Tanged Knives

Britain

124. Small tanged knife. Smooth green patina at top of blade.
L. 6.2, W. 9.6
Luce Bay, Wigtonshire. Bought by G.F. Lawrence at Fenton Sale, Nov. 1927.
927.3.17

125. Tanged knife. Slight median ridge along blade and two nicks on tang to secure hilt rivets. Surface rough; dark brown/green mottled patina.
L. 14.8, W. 37.3
Undley, Suffolk. Purchased by G.F. Lawrence at Fenton Sale, Nov. 1927.
927.3.14

126. Tanged knife. Slight median ridge along blade and waisted tang. Smooth, dark brown patina covers ca 1/5 of surface.
L. 14.7, W. 29.6
Undley, Suffolk. Purchased with no. 125 above, and listed under same number on invoice. Possibly associated finds?
927.3.15

127. Tanged knife with bevelled blade, flat tang and two nicks for hilt rivets. Condition excellent. Dark brown patina; smooth surface.
L. 11.4, W. 34.3
Derbyshire. Sturge colln.
918.33.8

Ireland

(Binchy 1967)

128. Tanged razor-knife. Broken across central tang hilt rivet hole. Low median ridge on blade defined by two shallow depressions. Surface smooth; pale green/brown mottled patina.
L. 7.2, W. 11.9
Found while digging potatoes, Ballymena, Co. Antrim. Ex Knowles colln., purchased through Fenton, Nov. 1924. Cf. Binchy 1967: figs. 2–3.
925.53.1

129. Tanged razor-knife. Low median ridge higher on one side to give plano-convex effect. Tang has central rivet hole. Dark brown patina with smooth surface. Excellent condition.
L. 11.7, W. 36.5
Co. Antrim. Sturge colln.
918.33.97

130. Tanged razor-knife. Bevelled blade and tang with central rivet hole. Ancient damage to tang butt. Surface smooth; patina pale brown.
L. 11.7, W. 22.0
Moorfields, Co. Antrim. Ex Knowles colln. Purchased with no. 128 above.
925.53.2

131. Tanged knife. Flat tang and blade with low median ridge. Central rivet notch near butt of tang. Condition excellent. Dark brown patina; smooth surface.
L. 14.6, W. 43.3
Near Armagh.
913x11.7

Socketed Knives

Britain

132. Blade of socketed knife. Sand and calcareous encrustation removed from one face to show smooth surface, with pale green patina.
L. 16.2, W. 53.4
Thames near Kew. Bought from G.F. Lawrence, Sept. 16, 1909.
909.67.3

Ireland

133. Socketed knife of Thorndon type. Peg-holes midway on socket; lightly faceted blade. Tip missing; blade nicked; socket squashed; otherwise condition good; surface smooth with dark brown patina.
L. 21.7, W. 158.6
Ireland.
913x11.9

134. Socketed knife of Thorndon type. Peg-holes towards socket mouth. Socket rim nicked; blade damaged. Brown patina; irregular surface.
L. 14.4, W. 72.7
Ireland.
913x11.11

Dirks and Rapiers

(Burgess 1968b; Trump 1962 and 1968)

The distinction between dirks and rapiers is that of Trump (1962:84), where a dirk is from 21.6 to 35.5 cm in length; rapiers must measure more than 35.5 cm from tip to butt.

Britain

135. Rapier fragment. Butt and upper blade only. One rivet in place, the other rivet hole damaged. Condition good; surface smooth; patina brown.
L. 10.5, W. 63.3
Thames near Maidenhead. Burgess (1968b) Group IV.
909.66.1

136. Dirk with two rivet notches in butt. Slender blade in excellent condition; surface smooth; dark brown/black patina.
L. 26.7, W. 84.5
Thames. Sturge colln. Burgess (1968b) Group IV.
918.33.7

137. Rapier with two rivet holes. Broad blade in excellent condition; surface smooth; patina dark brown.
L. 37.6, W. 202.1
River Ouse at Lakenheath. Bought by G.F. Lawrence at Fenton Sale, Nov. 1927. Another Fenland, more correctly Fen-edge, rapier (Trump 1968: Wandsworth-Chatteris Class). Burgess (1968b) Group IV.
927.3.4

138. Tang-hilted dirk with leaf-shaped blade and four rivet notches on hilt. Condition excellent; blade smoother than tang; very dark brown patina.
L. 23.2, W. 103.2
Thames at Battersea. Bought from G.F. Lawrence, Nov. 15, 1910.
910.153.2

139. Tang-hilted dirk with two rivet notches. Metal between the two notches deeply grooved on both faces in antiquity. Thin, sharp butt. Blade slightly bent. Uneven surface; patina pale green.
L. 27.4, W. 111.2
Thames. Sturge colln. Burgess (1968b) Group II.
918.33.5

140. Rapier blade. Broken in antiquity. Cutting edges severely damaged, otherwise surface smooth; patina whitish green.
L. 26.5, W. 207.4
Thames. Burgess (1968b) Group IV.
909.66.3

IRELAND

141. Tanged dirk with (?) two rivet notches. Cruciform butt characteristic of Cutts class (Trump 1962), but blade is narrow, not leaf shaped. Base of tang broken, cutting edge nicked and bent in antiquity. Surface texture sandy; patina pale green and brown.
L. 22.2, W. 66.2
Near Armagh.
913x11.5

Swords

BRITAIN

(Cowen 1951 and 1967)

142. Leaf-shaped sword of Ewart Park type. Lightly curved ricasso and three surviving rivet holes. Part of tang missing and tip of blade bent. Condition excellent with smooth surface and thin dark brown patina.
L. 48.8, W. 546.9
Thames at Shepperton. Bought from G.F. Lawrence, Jul. 19, 1927.
927.28.102

143. Tip portion of blade from leaf-shaped sword. Probably broken in antiquity. Light sandy concretion on both faces. Patina pale brown on one side, pale green on the other; smooth surface.
L. 18.5, W. 128.9
Thames near Reading. Bought from L. Treacher, Twyford.
907.16.5

144. Leaf-shaped sword of Ewart Park type. Three rivet holes and splayed pommel tang. Ricasso slightly curved with rounded profile. Condition excellent; surface quite smooth; patina dark brown.
L. 52.3, W. 532.5
Thames, near Down Place, above Windsor. Bought from L. Treacher, Twyford.
908.32.1

145. Leaf-shaped sword of Ewart Park type. Widely splayed pommel tang. Six rivet holes with two rivets in place. Low hilt flanges. Slightly curved ricasso of rounded profile. Condition excellent; smooth surface with very dark brown/black patina.
L. 58.4, W. 622.6
Thames, near London. Ex colln. C. Roach Smith, Apr. 1885. Bought from W. Talbot-Ready, Sept. 1907.
909.14.3

146. Leaf-shaped sword of Wilburton type. Splayed pommel tang. Three rivet slots. Curved ricasso of rounded profile. Tip missing and blade broken. Surface smooth; whitish green/brown patina with calcareous encrustation towards butt.
L. 62.1, W. 789.1
Thames at Maidenhead. Bought from L. Treacher, Twyford.
908.32.3

147. Leaf-shaped sword of Wilburton type. Splayed pommel tang. Three rivet slots with two rivets surviving. Ricasso curved, with rounded profile. Blade tip missing. Surface much pitted, except for hilt and top of blade where light brown/green patina intact.
L. 59.4, W. 579.8
Thames. Bought from L. Treacher, Twyford.
908.32.2

Ireland

(Eogan 1965: 161–162)

148. Leaf-shaped sword of Ewart Park type. Hilt tang broken off near butt. Three rivet holes and short rivet slot. Hilt tang slightly damaged by the attachment in 1798 of a crude leather handle secured by rough iron rivets (since removed). Condition excellent; surface smooth with rich red brown patina.
L. 62.6, W. 831.8
Taken at New Ross, Co. Wexford in the rebellion of 1798. Probably from SE Ireland. Bought from Fenton and Sons, Aug. 1909.
909.68.1

149. Leaf-shaped sword of Ballintober type. Hilt tang damaged; upper part of blade battered and nicked. Two rivet holes with rivets. Black stain on butt left by wooden or leather hilt. Condition excellent; surface smooth; patina brown.
L. 50.2, W. 509.3
Near Armagh. From St. Columba's College.
913x11.1

150. Leaf-shaped sword of Ewart Park type. Flared pommel tang and four rivet holes; very low hilt flanges. Short ricasso of square section. Condition excellent with smooth surface; light brown patina.
L. 44.4, W. 329.3
Dunloy, W. Antrim. From St. Columba's College.
913x11.3

151. Leaf-shaped sword of Ewart Park type. The hilt tang above the shoulders is missing. Two shoulder rivet holes. Blade broken at narrowest point and repaired. Condition good; slightly rough surface; patina brown.
L. 45.9, W. 378.1
Enniskillen, Co. Fermanagh. From St. Columba's College.
913x11.2

152. Lower portion of leaf-shaped sword blade. Tip missing. Part nearest butt modified (? in antiquity) by boring hole and scouring top of blade. Cutting edge sharpened recently. Smooth surface; patina dark brown.
L. 27.3, W. 235.6
Ireland. Sturge colln.
918.33.96

153. Fragment of leaf-shaped sword. Upper hilt tang and most of blade missing. Five rivet holes, of which the upper three are blocked with small rivets cut off flush with the surface. Cutting edges deliberately blunted. No ricasso. Condition good; slightly rough surface; patina dark brown.
L. 20.6, W. 272.7
Probably Irish. From St. Columba's College. Hallstatt type (cf. Cowen 1967: 391 ff.).
913x11.6

Miscellaneous

Britain

154. Small tanged awl.
L. 4.2, W. 1.6
Icklingham, Suffolk. Bought from G.F. Lawrence, Jan. 7, 1927.
927.28.101

155. Looped socketed chisel. Flared cutting edge and three rounded ribs around rim. Centre of cutting edge damaged (? through use). Condition good; surface slightly rough-cast; patina dark brown.
L. 10.9, W. 157.8
Suffolk. Bought from W.S. Simpson.
908x34.6

156. Unlooped socketed gouge. Plain. Pale green smooth patina survives in places.
L. 6.3, W. 55.6
Found at Tillingham, Essex, 1827. Sturge colln.
918.33.4

157. Part of tubular ferrule. Filled with solid mass of sand, gravel, and shell. Broken at both ends. Exterior cleaned—brown and smooth.
L. 9.2
Thames near Reading. Bought from L. Treacher, Twyford.
907.16.24

158. Unlooped socketed chisel. Plain. Socket extends 3.2 cm below rim. Sandy concretion on upper part; surface slightly rough; patina green.
L. 12.7, W. 85.6
Norfolk. Sturge colln.
918.33.3

159. Socketed sickle with peg-holes. Two slight ribs around socket mouth. Condition excellent; smooth surface; brown patina.
L. (socket to heel) 7.0, W. 97.0
Thames at Hampton Court. Bought from G.F. Lawrence Oct. 26, 1933. Belongs to Fox's (1939) Group I—Thames series.
933.10.34

160. Three-linked snaffle-bit. Made of bronze throughout. Condition excellent. Patina brown on rings and outer links, but red brown on central link which is rougher than the rest.
L. 24.5, W. 204.2
Dredged from the river Conway, Wales, 1842. Ex colln. G.F. Lawrence (bought from him Oct. 9, 1935). An Iron Age type (Ward-Perkins 1939).
935.12.50

161. Tongue-shaped chape. Brown/pale brown patina; smooth surface.
L. 19.4, W. 106.4
Thames at Sion Reach. Bought from G.F. Lawrence, Oct. 6, 1928.
928.35.6

162. Razor. Blade of oval plan with circular central perforation and short ringed handle. Condition excellent; very dark brown patina.
L. 5.85, W. 14.1
From a barrow in Cornwall. Sturge colln. A Hallstatt type.
918.33.6

163. Circular shield of Coles's (1962) Harlech type. The lower two tabs, rivets, and rivet holes on either side are modern additions. Both handle and tab rivets have domed heads. The condition of the metal is good, as conservation has removed the patina. The ten ribs have been beaten out in the manner suggested by Coles (1962:181 ff.). The thickness of the sheet metal varies from 0.8 mm to 1.0 mm; the diameter of the shield is 70.0 cm and the diameter of the central boss is 14.5 cm. The total weight, including modern additions, is 2.75 kg.
A letter from Dr. Corner to Dr. Currelly states that the shield was dredged from the river Lea near Ponders End. It would appear that the damage was caused by the steam navvy used in the work. Dr. Corner bought the shield from the dredging contractors and the ROM acquired it in 1935.
Publications: Smith 1919; Coles 1962: no. 5, p. 190.
935.58

Ireland

164. Solid-cast ring with lateral buffer-shaped perforated projections. Condition excellent; smooth surface; dark brown/green patina.
L. (across terminals) 7.4, Dia. 6.4, W. 37.1
Near Limerick, 1854. Sturge colln. Possibly a type of horse trapping, of a sort peculiar to Ireland. Cf. Eogan 1964: fig. 16, no. 13; Evans 1881: fig. 497.
918.33.94

165. Solid casting of three concentric rings, the exterior of the largest ring being decorated with two semicircular arrangements of ten and nine cast-on loops. The two 1.0 cm gaps between both sets of loops show signs of wear or deliberate smoothing. Condition excellent; surface smooth; patina dark brown.
Dia. 9.6 (across loops) to 8.6 (outer solid ring); W. 528.9
Near Limerick. Sturge (ex Bateman colln.). Cf. Wilde 1857: fig. 491.
918.33.98

166. Solid bronze ring with tubular lateral projection. The tubular projection is decorated at both ends with three closely grouped, low raised bands. The tube is perforated just below the three bands around its mouth by two pin or peg-holes which were cut, or finished, after casting.
Dia. 4.7, L. 9.5, W. 245.2
Ireland. Sturge (ex Bateman colln.). Possibly a terret?
918.33.93

167. Crescent-headed linchpin of (Ward-Perkins 1939) Type II. Foot of shaft has broken across pinhole near base. Low median ridge on one face of crescent head only. Upper surface of head decorated with spaced zones of transverse lines. Sandy surface texture; pitted on back; light green patina.
L. 11.1, W. 315.8
Dunmore, Co. Galway. Sturge coll. Iron Age.
918.33.95

168. Halberd with asymmetrical blade and low rounded median rib. Three rivet holes in butt.

163a Late Bronze Age shield dredged from the river Lea, front view. Scale in centimetres. Photo by L. Warren, ROM.

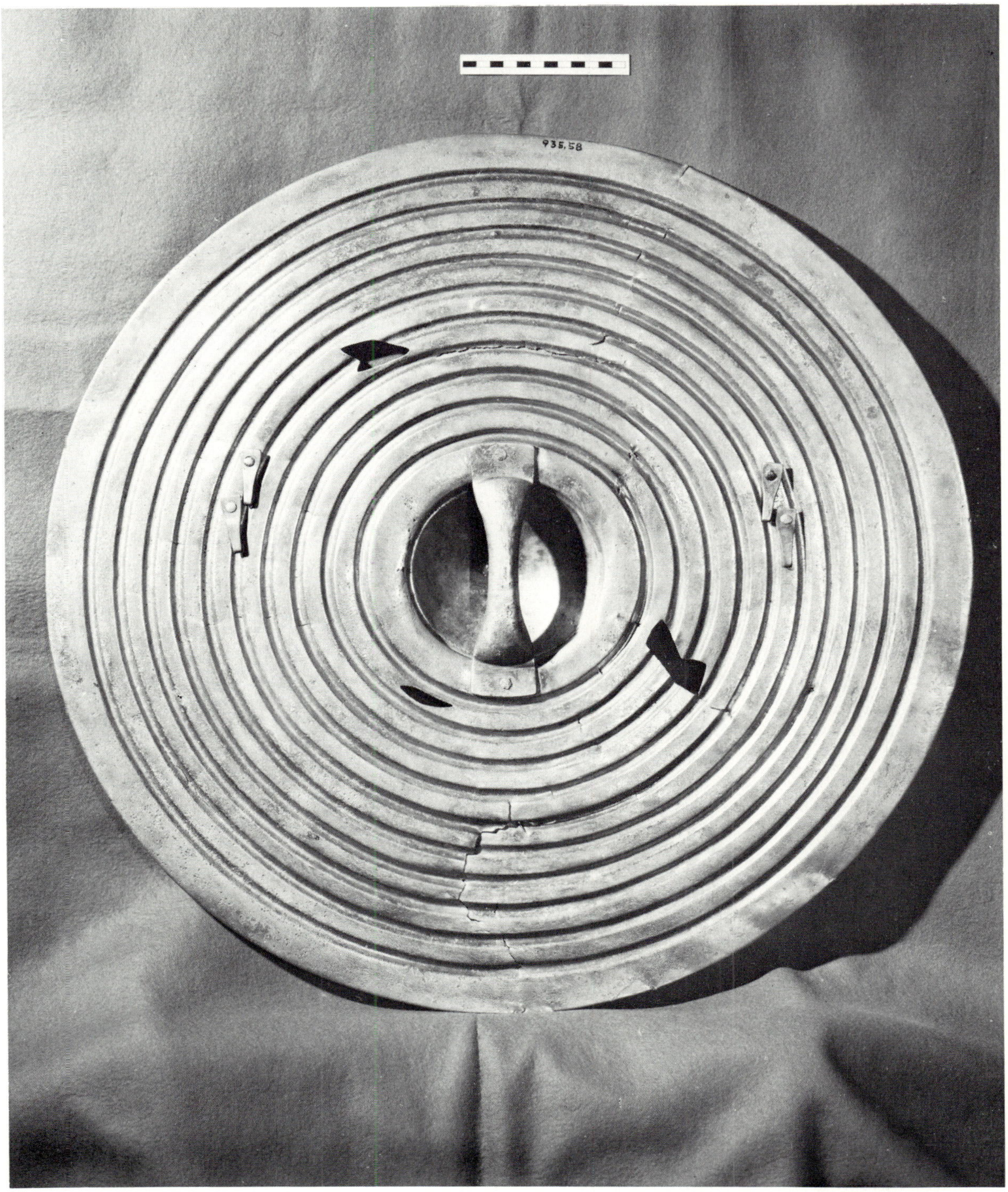

163b Late Bronze Age shield dredged from the river Lea, rear view. Scale in centimetres. Photo by L. Warren, ROM.

Some damage to butt, otherwise condition excellent; surface slightly rough; patina dark brown.
L. 24.8, W. 259.5
Castletown, Berehaven, Co. Cork. Day Sale, 1913. Listed in Harbison (1969b:41, no. 197) as "present whereabouts unknown".
913x11.4

Hoards

THAMES, NEAR READING

(nos. 89, 90, 91, and 94)

This hoard has not been separated from the other pieces in the collection because the association seems doubtful.

WIMBLEDON, SURREY

(nos. 169–173)

This hoard, or part of a hoard, consisting of four Late Bronze Age socketed axes and one winged axe, was found at Wimbledon and bought by the ROM from G.F. Lawrence, June 26, 1930. Although the individual pieces differ very slightly in their condition and preservation, they all carry small traces of the same pale sandy soil.

169. Looped socketed axe. Single horizontal rib 1.6 cm below socket rim. Two pronounced internal vertical ribs. Blunt cutting edge. High casting seams. Dark green/brown smooth patina covers ca ½ the surface.
L. 10.75, W. 301.7
Wimbledon hoard (see above).
930.83.17

170. Faceted looped socketed axe. Single small horizontal rib 2.0 cm below socket rim. Slightly flattened loop. Cutting edge damaged. Socket wall cracked in two places opposite loop. Dark green/brown smooth patina covers ca ¾ of the surface.
L. 10.4, W. 180.0
Wimbledon hoard (see above).
930.83.20

171. Looped winged axe. Wings cracked and damaged by hammering. Cutting edge damaged. Smooth brown patina covers ca ¾ of the surface.
L. 14.05, W. 319.2
Wimbledon hoard (see above).
930.83.16

172. Looped socketed axe. Single horizontal rib 1.4 cm below socket rim. Casting flaw opposite loop. Cutting edge sharp. Dark green/brown smooth patina covers ca 4/5 of the surface.
L. 9.6, W. 174.2
Wimbledon hoard (see above).
930.83.18

173. Looped socketed axe. Single horizontal rib 1.5 cm below socket rim. Cutting edge damaged. Dark green/brown smooth patina covers ca ¼ of the surface.
L. 6.85, W. 124.0
Wimbledon hoard (see above).
930.83.19

HARROGATE, YORKSHIRE

(nos. 174–176)

The three pieces of this Late Bronze Age hoard (a leaf-shaped Ewart Park sword, a subtriangular basal-looped socketed spearhead, and a leaf-shaped socketed spearhead with gold band decoration) were found as a group near Harrogate (cf. Coles 1964:193 n. 7). They were originally acquired, for his personal collection, by the dealer S.G. Fenton, proprietor of "The Old Curiosity Shop", Leicester Square, London. Upon his death, his brother W.H. Fenton, senior partner of the prestigious antiquities dealers, Fenton and Sons, sold his remaining stock and private collection at Sotheby's, where they were bought for the ROM by another dealer, G.F. Lawrence, in May 1927.

This important hoard has recently been fully discussed by Coles (1964), in relation to a very similar personal hoard from Pyotdykes, Angus. The spearhead with gold band decoration (no. 174) was further discussed by Coles in 1971.

174. Leaf-shaped socketed spearhead with peg-holes and gold band decoration around the socket. The gold band (Coles 1964:pl. xix; 1971:pl. xxii) is located immediately above the socket mouth and is 1.9 cm wide. The decoration consists of five zones:

three of horizontal lines bounding and separating two of interlocking pendant and erect triangles filled with alternate hatching. This pattern was cut into the bronze, after which the thin gold sheet was applied. A sharp tool was then used to remove the gold from the grooves. The edges of the blade have been nicked and the socket mouth is slightly damaged. There are traces of a reddish brown patina which was not removed in the cleaning process.
L. 41.9, W. 479.4
Harrogate hoard (see above).
927.3.9

175. Leaf-shaped sword of Ewart Park type. Hilt tang damaged. Four hilt rivet holes survive. Lightly curved ricasso of rounded profile. Blade tip slightly damaged and bent. Condition excellent, but cleaned and polished.
L. 56.5, W. 689.8
Harrogate hoard (see above).
927.3.10

176. Basal-looped socketed spearhead with straight-based blade of subtriangular form and peg-holes. Two ribs, parallel to the median ridge, spring from basal loop fillets. The socket narrows very slightly towards the mouth, where the metal is thin (1.3 mm). Condition excellent, apart from a few nicks on the blade. Cleaned.
L. 61.9, W. 847.5
Harrogate hoard (see above).
927.3.8

Ornaments

Britain

177. Cast torque with plain hook terminals and mock torsion decoration. The body of the torque has broken opposite the terminals, but otherwise the condition is excellent and the patina pale green.
Dia. 21.4, W. 480.3
Sussex. Gift of Mr. Thomas Sutton, Dec. 18, 1922.
Cf. Smith 1959: fig. 2; Hawkes 1942: pl. vi.
922.37.3

178. Armlet or bracelet. Decorated with two discontinuous parallel lines, 3 mm apart, ca 1 mm from left-hand edge. Sand encrustation on interior. Light brown patina; surface smooth.
Dia. 6.9–6.5, W. 58.9
Thames near Maidenhead. From L. Treacher, Twyford. Date uncertain.
907.16.25

179–181. Three fragments of ribbed bracelet. Bent and damaged.
Thames at Maidenhead. From L. Treacher, Twyford.
907.16.30

182. Bracelet fragment. Convex transverse section. Rough surface; light brown patina.
Thames near Maidenhead. From L. Treacher, Twyford.
907.16.30

Ireland

The two long pins, nos. 183 and 184, were purchased by the Museum in 1918 with the Sturge collection, but unlike other Sturge pieces their provenance, Tara, Co. Meath, seems suspect. "Tara" was often used to lend spurious authenticity to objects held by dealers which may not even have originated in Ireland.

The shape and length of the pins find no ready parallels in prehistoric Ireland (G. Eogan, pers. comm.), nor are pins of this type found in prehistoric Britain, where the nearest parallels would appear to be pins of Hawkes's (1942) Picardy type. Despite a superficial resemblance, our examples are both too long and lack the diagnostic perforated swelling of the true Picardy pin. However, whether they were found in Ireland or not, the metal composition of the pins is very similar (see below) and the low concentrations of both lead and zinc tend to argue in favour of a Bronze Age date.

183. Very long pin with expanded flat head. Decorated area on upper shank (5.5 cm wide) of five zones: three of horizontal grooves separate and bound two of erect triangles filled with short horizontal lines. Condition excellent; surface smooth; patina dark brown.

L. 45.3, W. 67.4
For provenance see above.
918.33.92

184. Very long pin with flattened biconical head. Decorated area on upper shank (5.0 cm wide) of five zones: three of widely spaced horizontal grooves separate and bound two of closely spaced, lighter, horizontal grooves. Condition identical to no. 183.
L. 47.6, W. 70.6
For provenance see above.
918.33.91

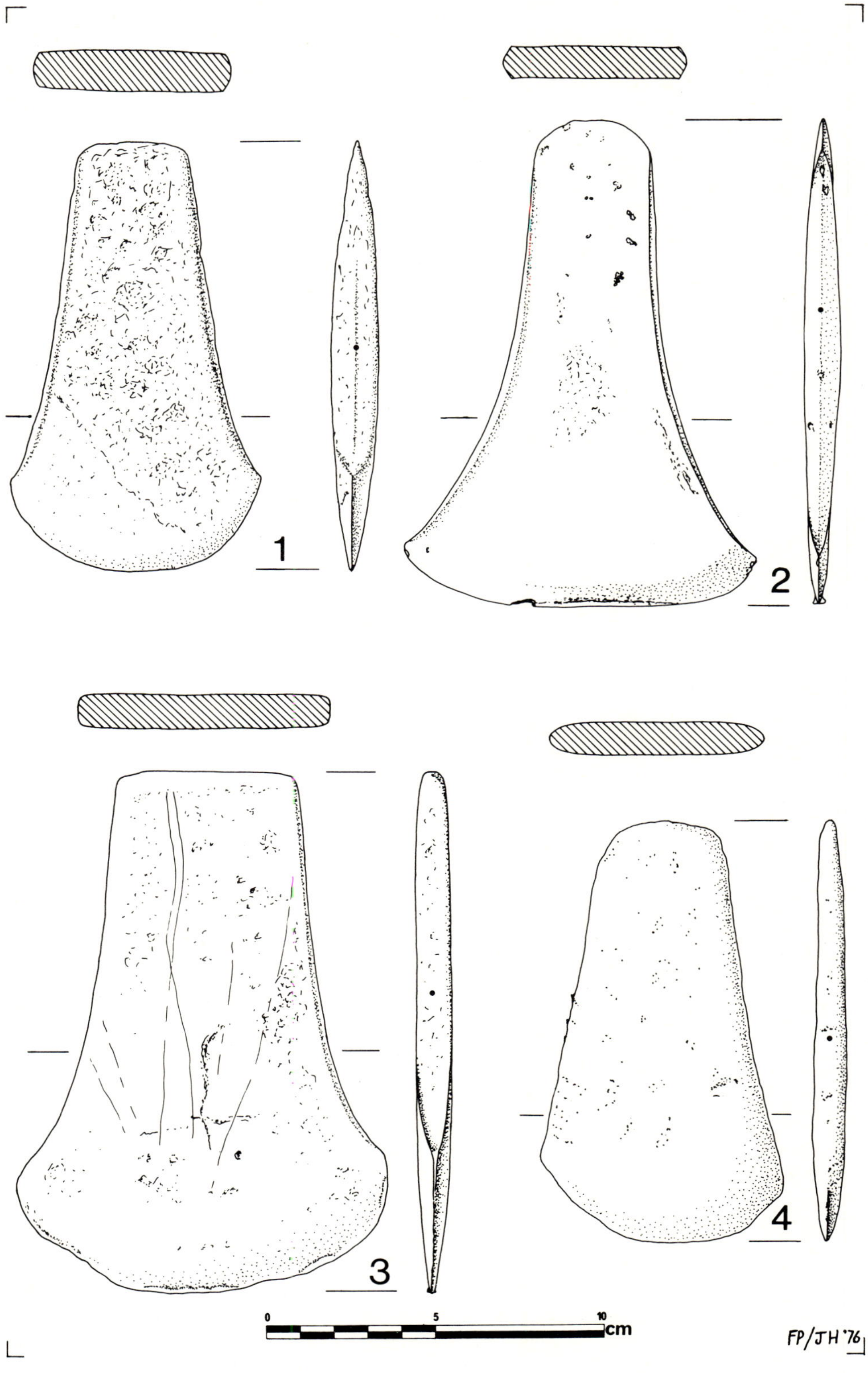
1
2
3
4
0
5
10
cm
FP/JH '76

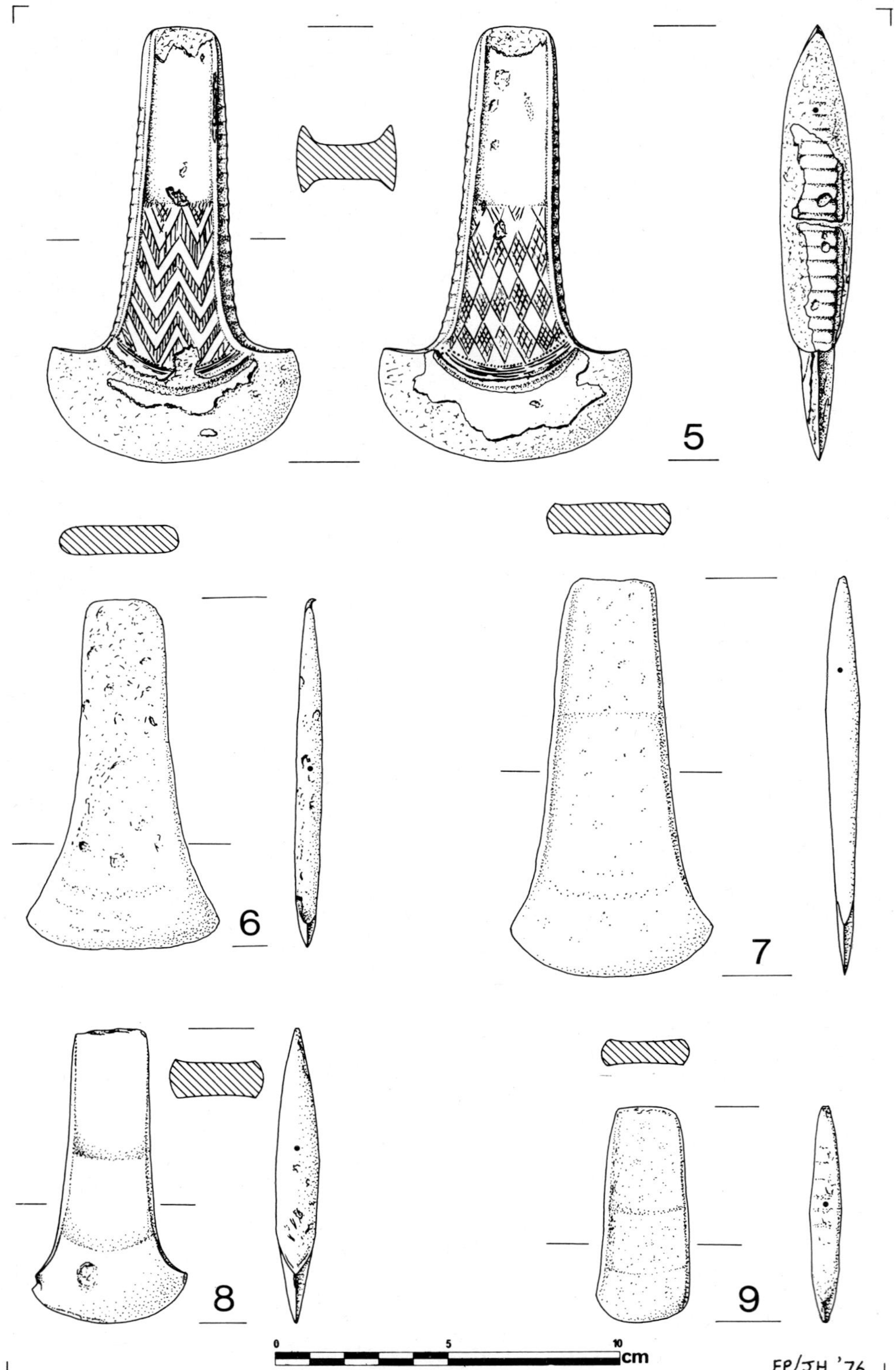
5
6
7
8
9
0
5
10
cm
FP/JH '76

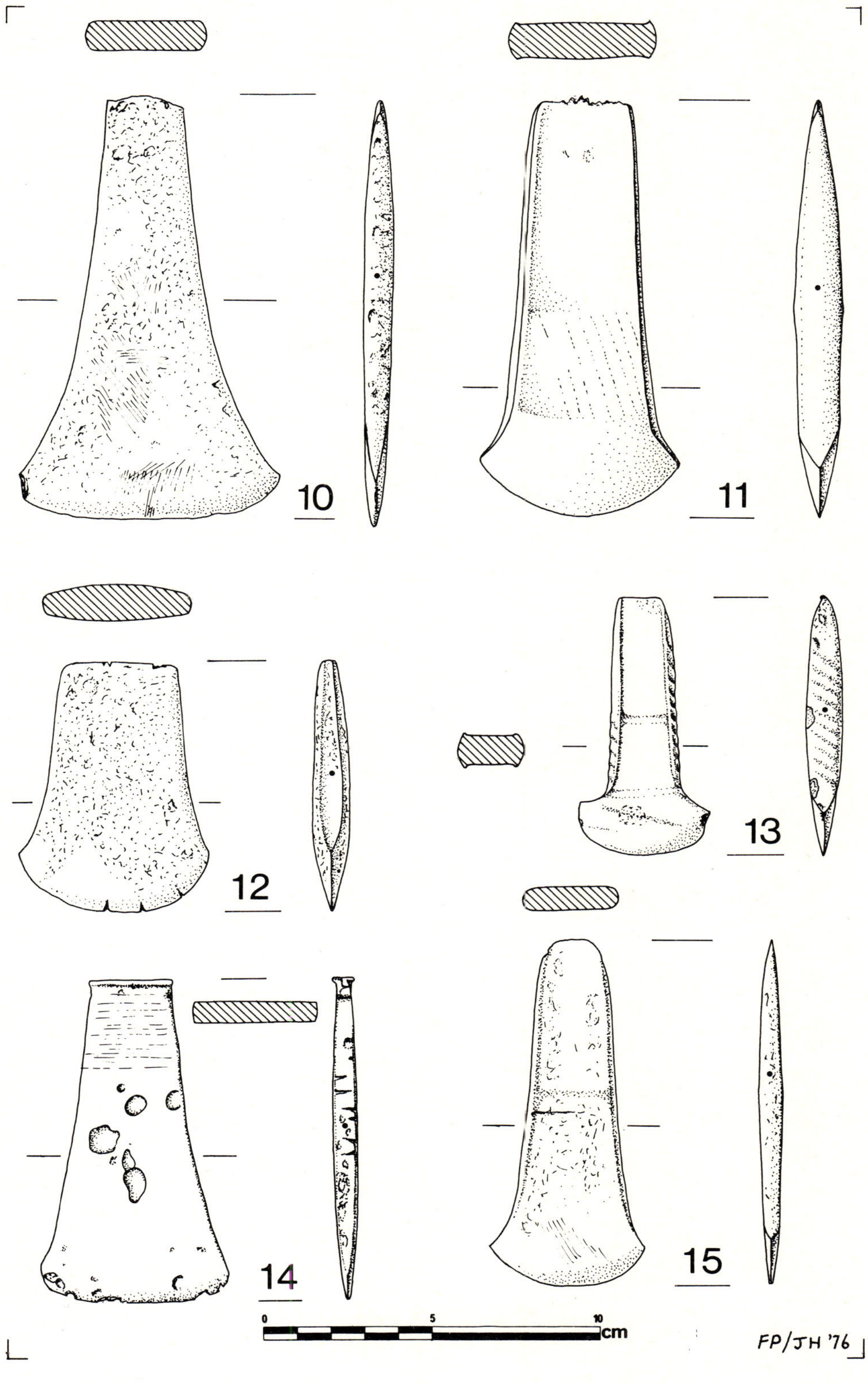
10
11
12
13
14
15
0
5
10
cm
FP/JH '76

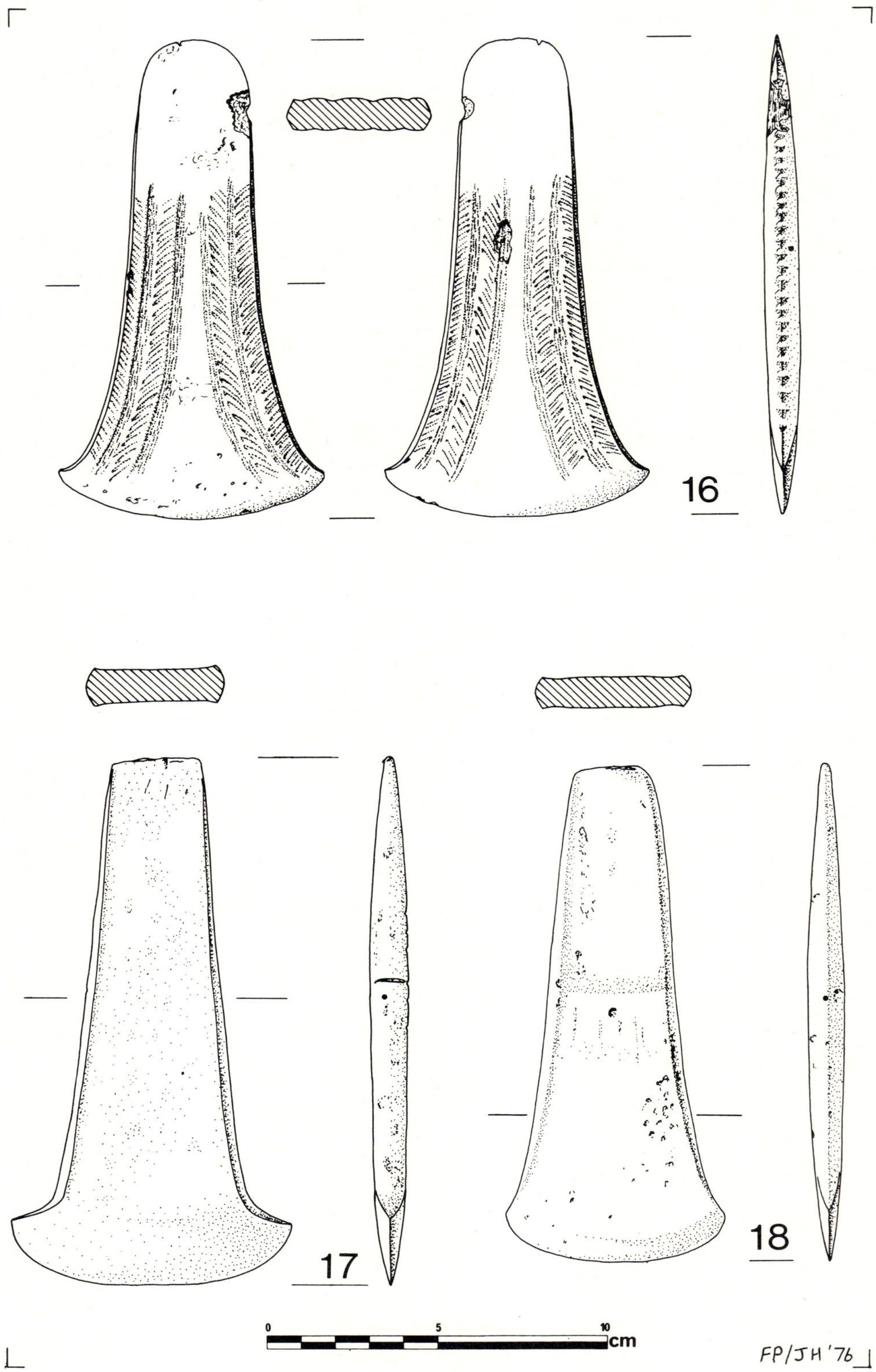
16
17
18
0
5
10
cm
FP/JH '76

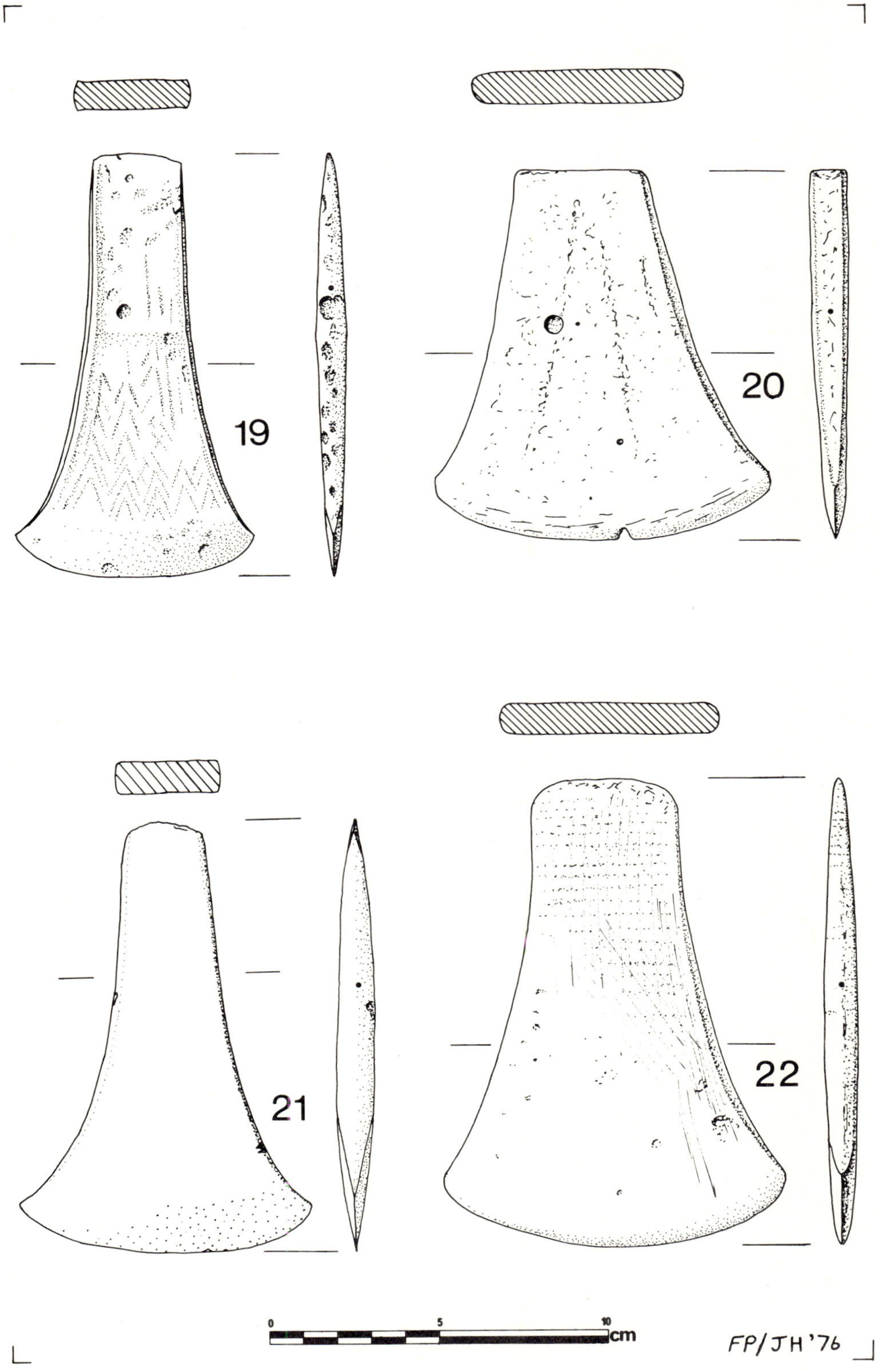
19
20
21
22
0
5
10
cm
FP/JH'76

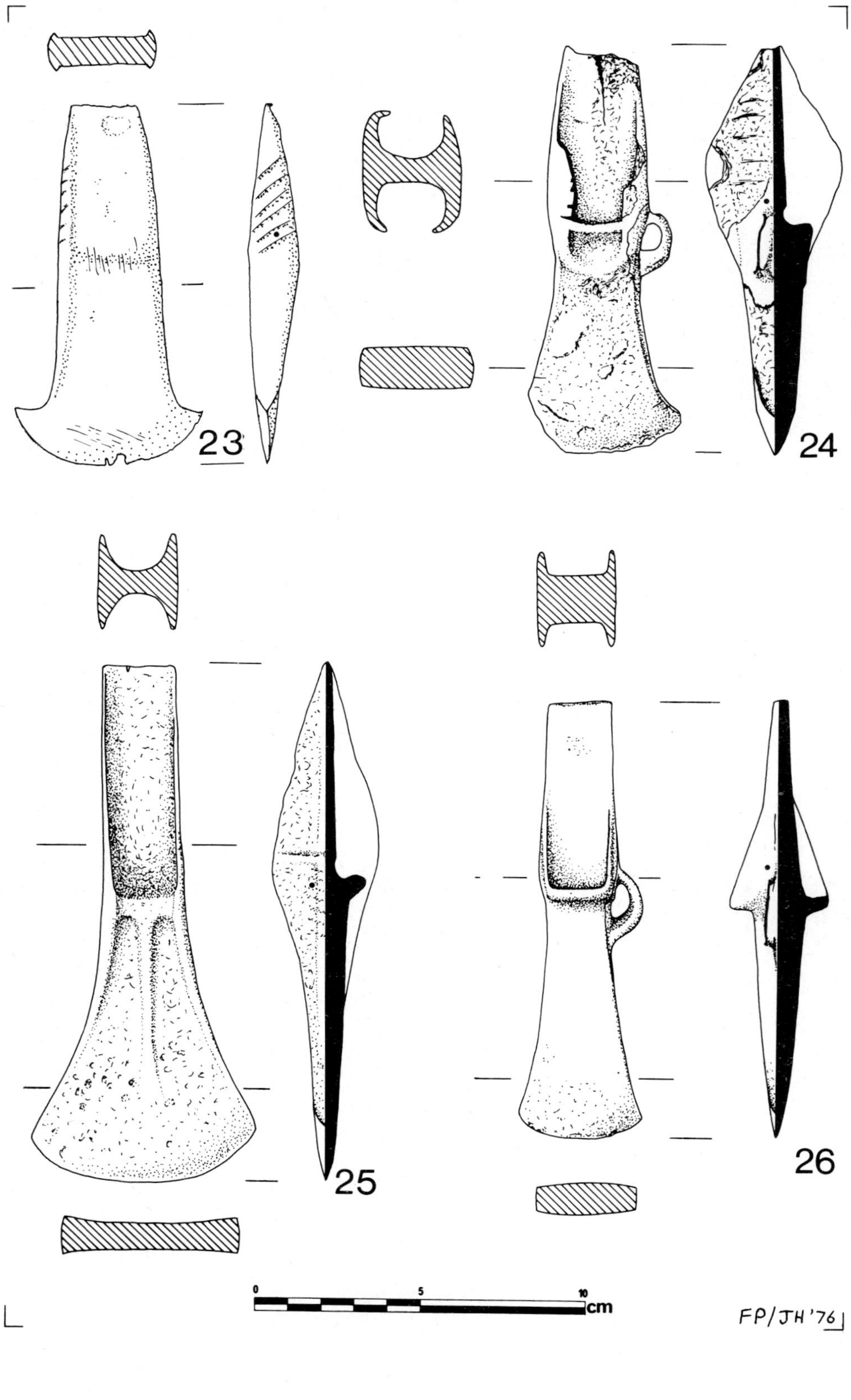
23
24
25
26
0
5
10
cm
FP/JH '76

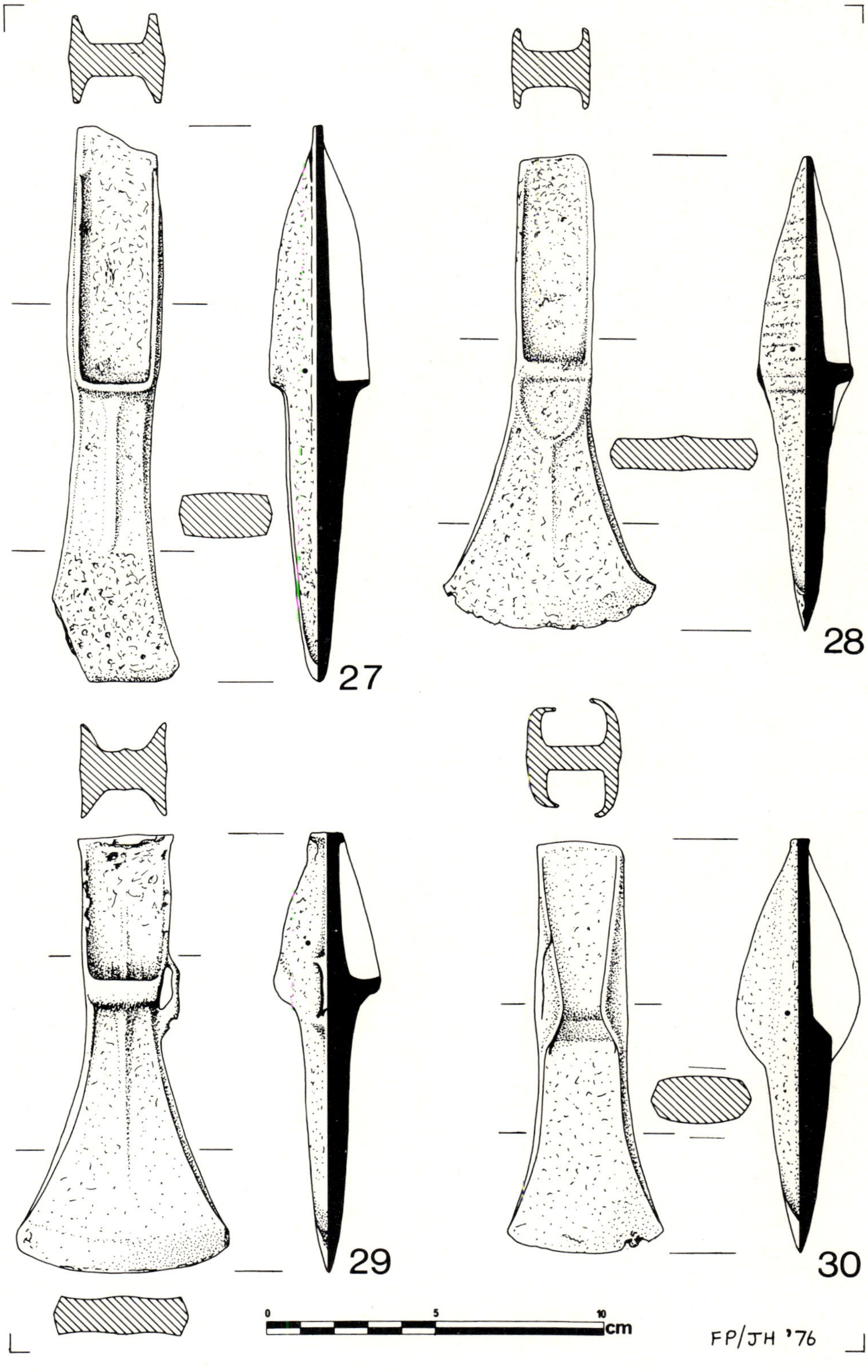
27
28
29
30
0
5
10
cm
FP/JH '76

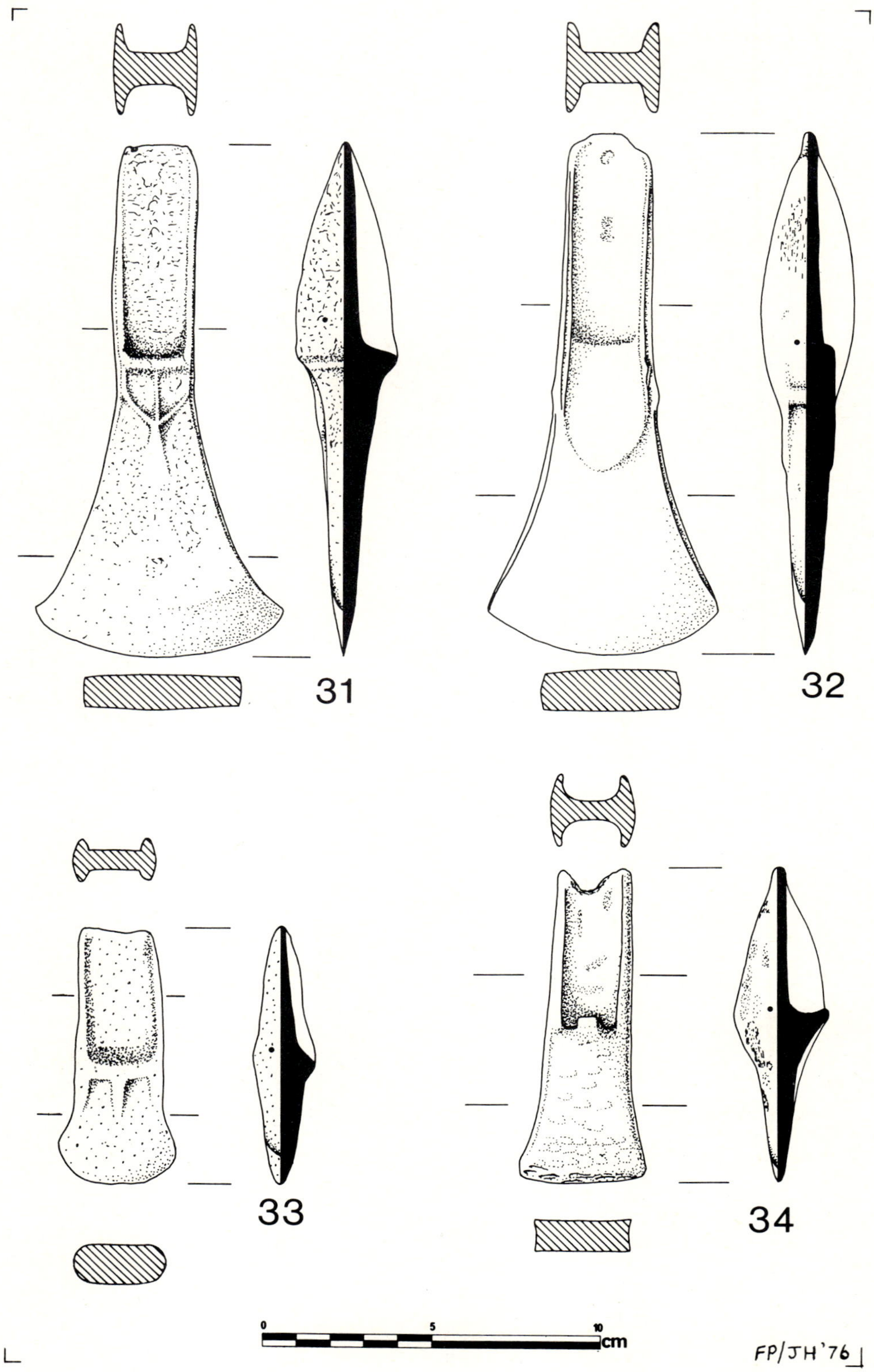
31
32
33
34
0
5
10
cm
FP/JH'76

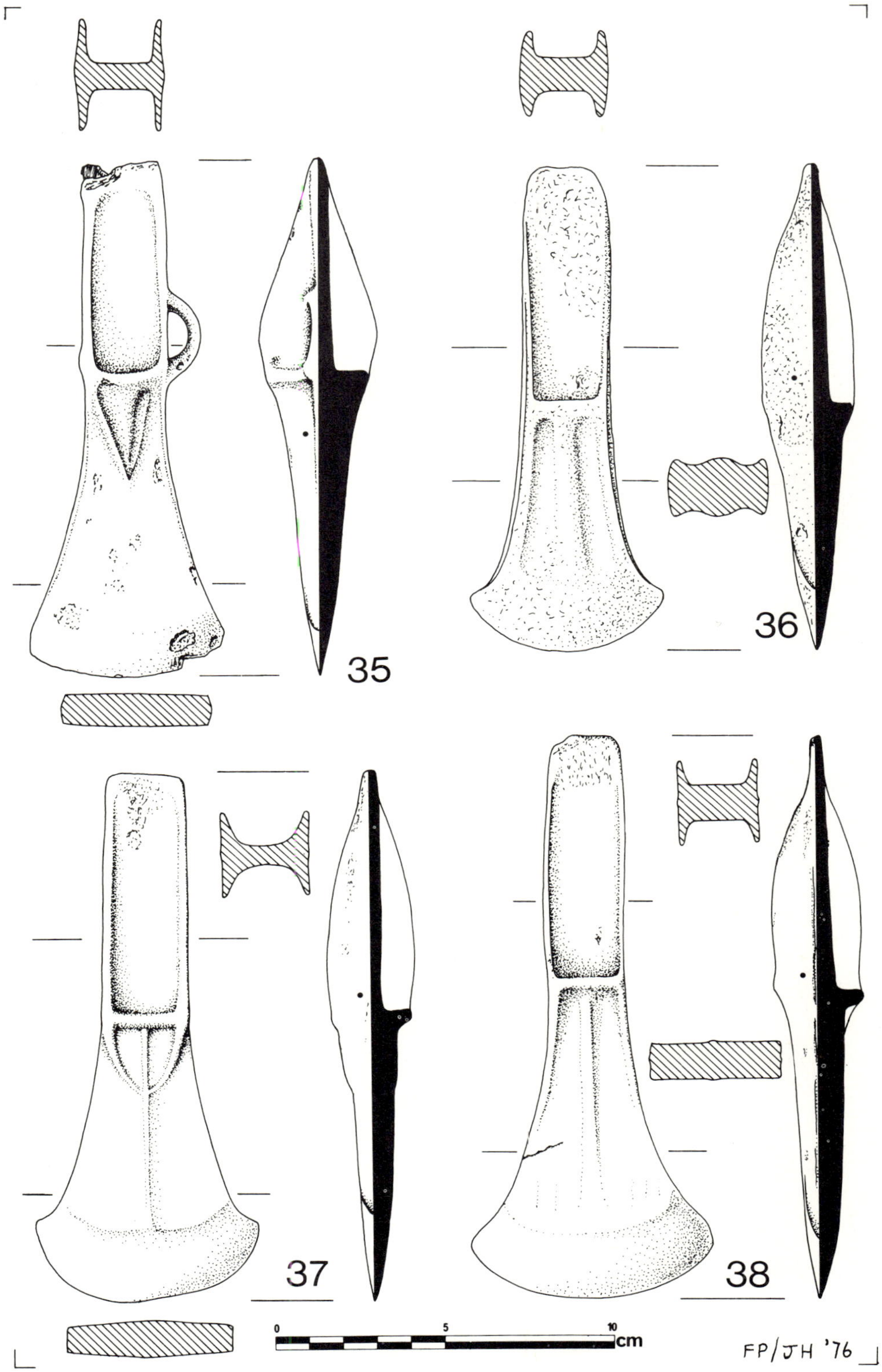
35
36
37
38
0
5
10
cm
FP/JH '76

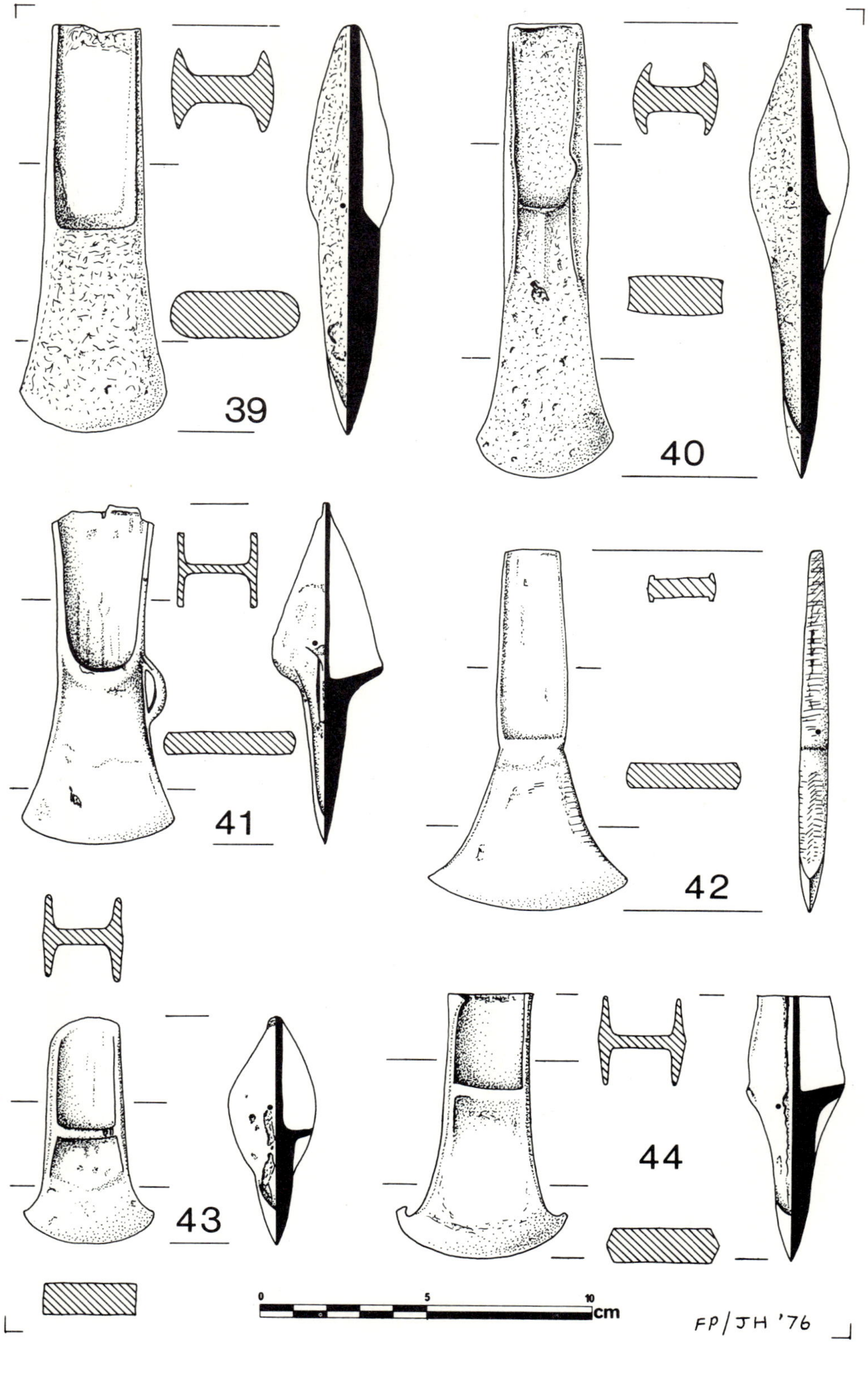
39
40
41
42
43
44
0
5
10
cm
FP/JH '76

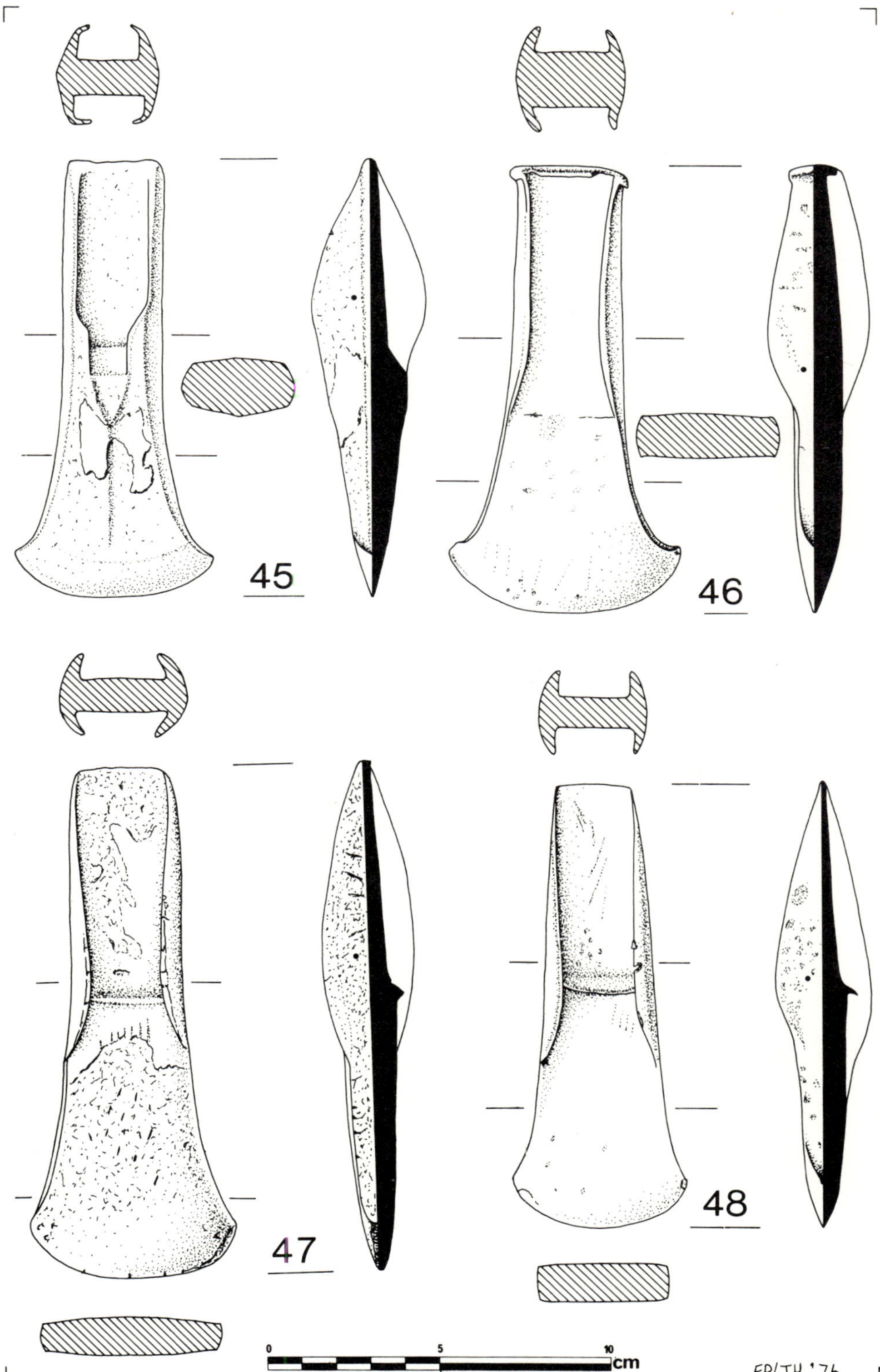
45
46
47
48
0
5
10
cm
FP/JH '76

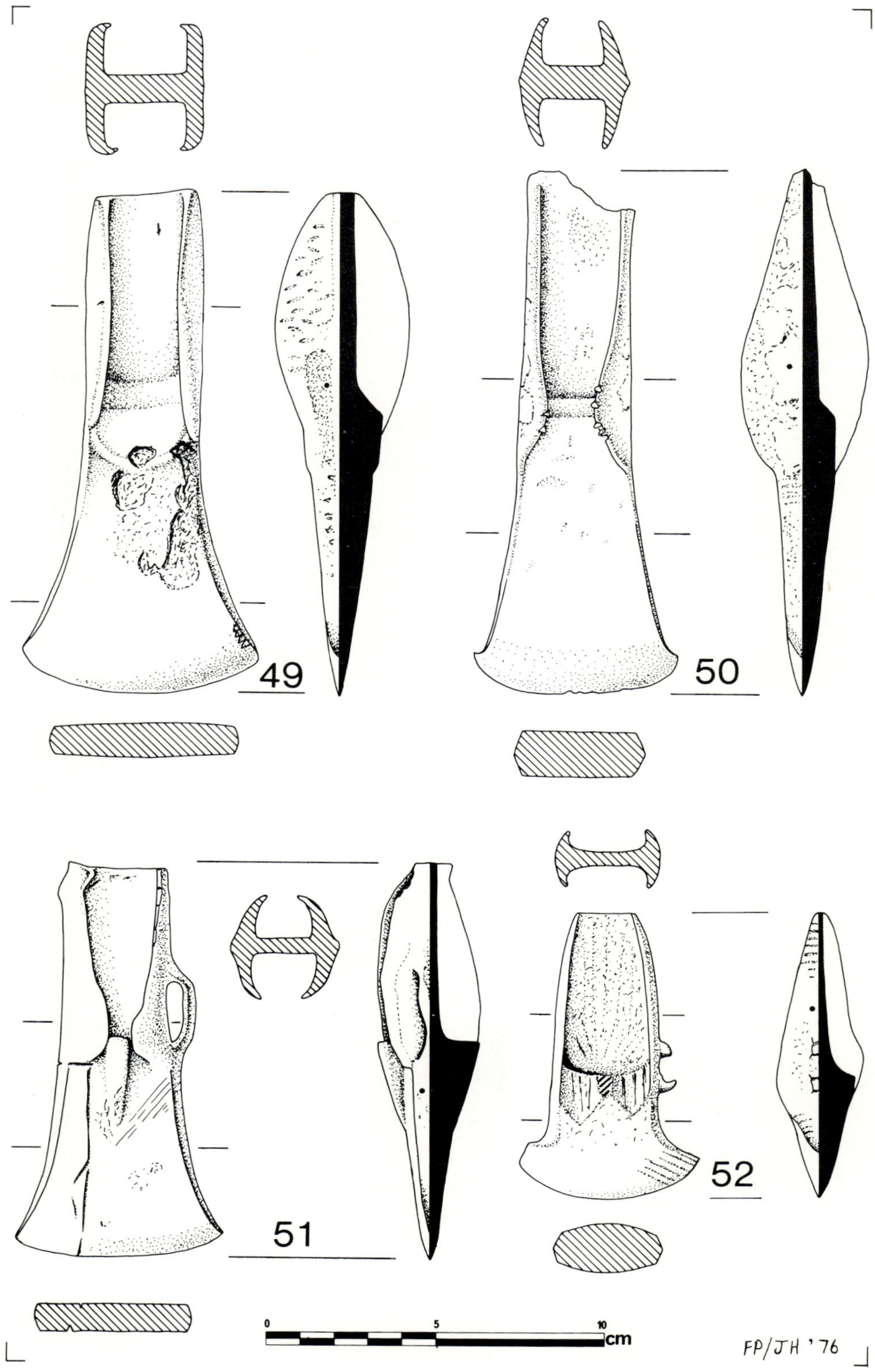
49
50
51
52
0
5
10
cm
FP/JH '76

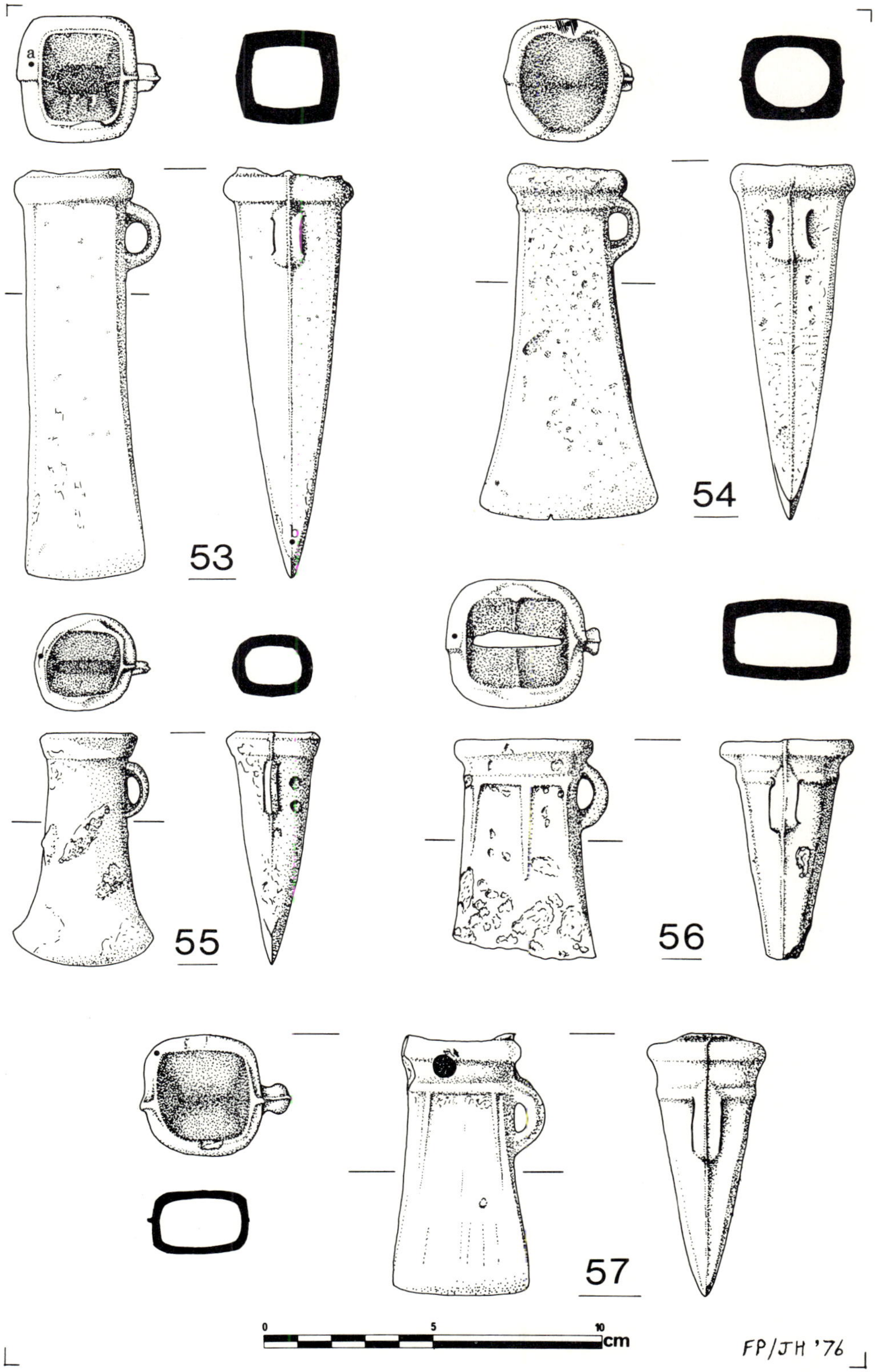
53
54
55
56
57
0
5
10
cm
FP/JH '76

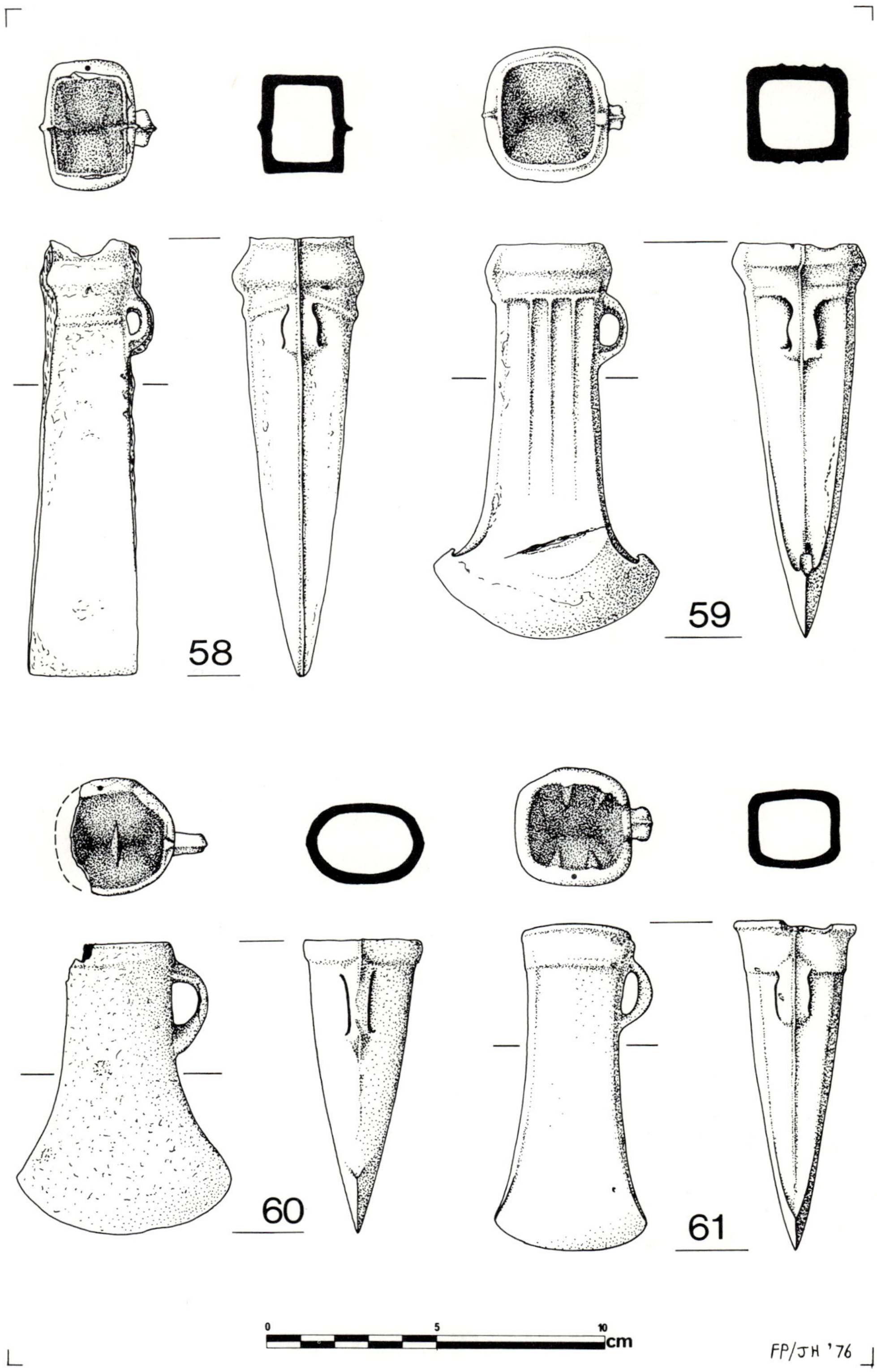
58
59
60
61
0
5
10
cm
FP/JH '76

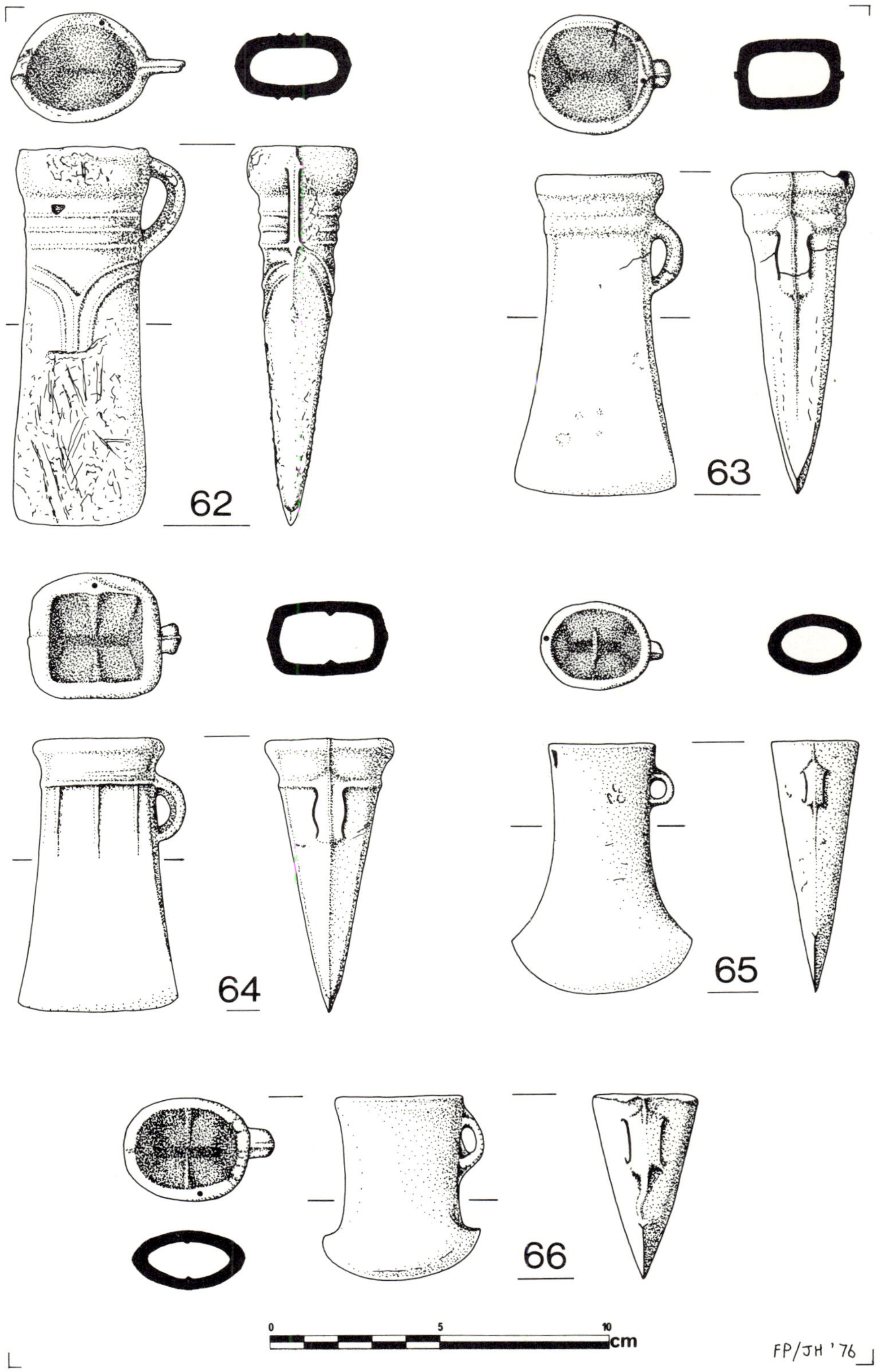
62
63
64
65
66
0
5
10
cm
FP/JH '76

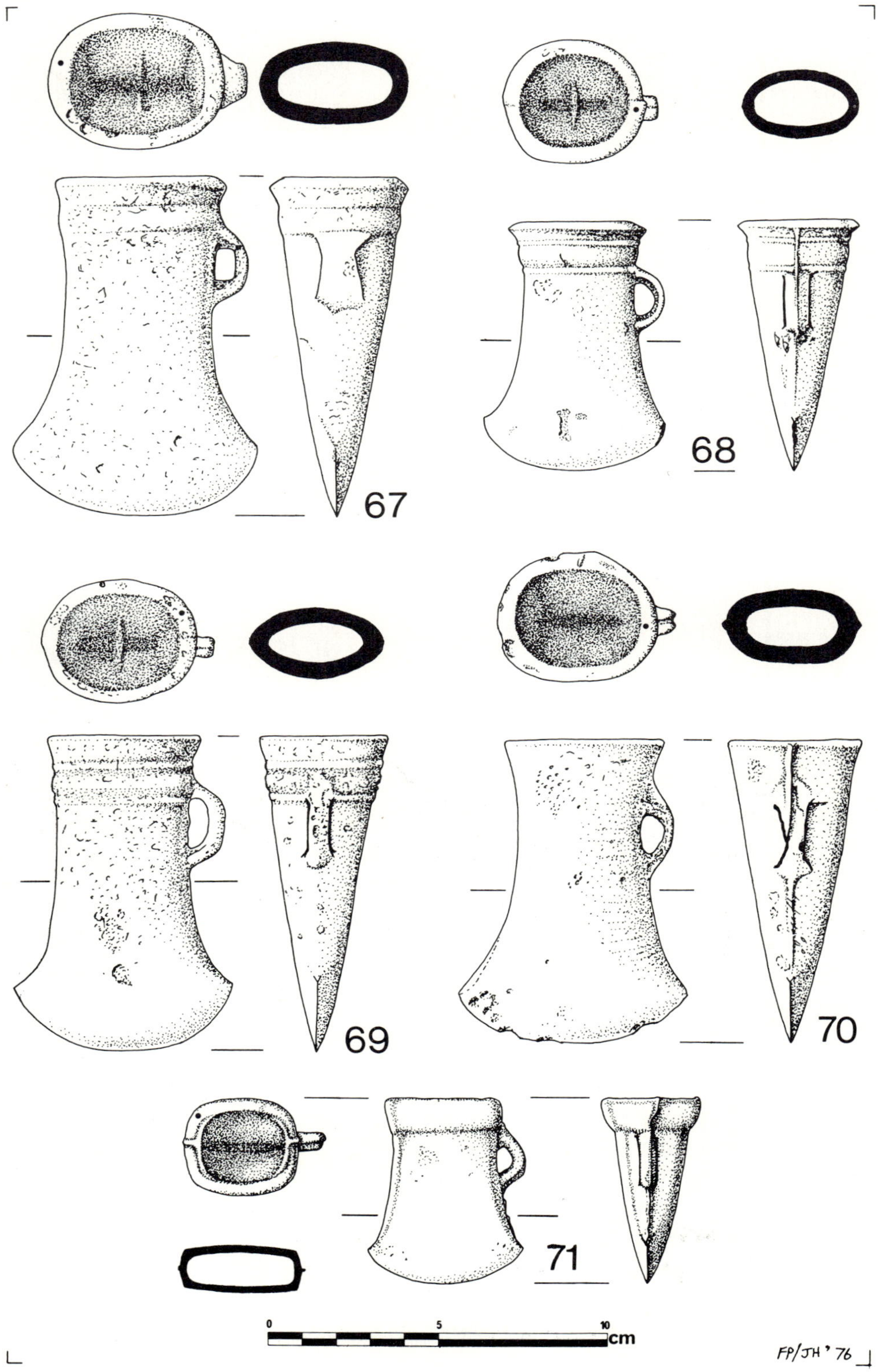

67
68
69
70
71
0
5
10
cm
FP/JH '76

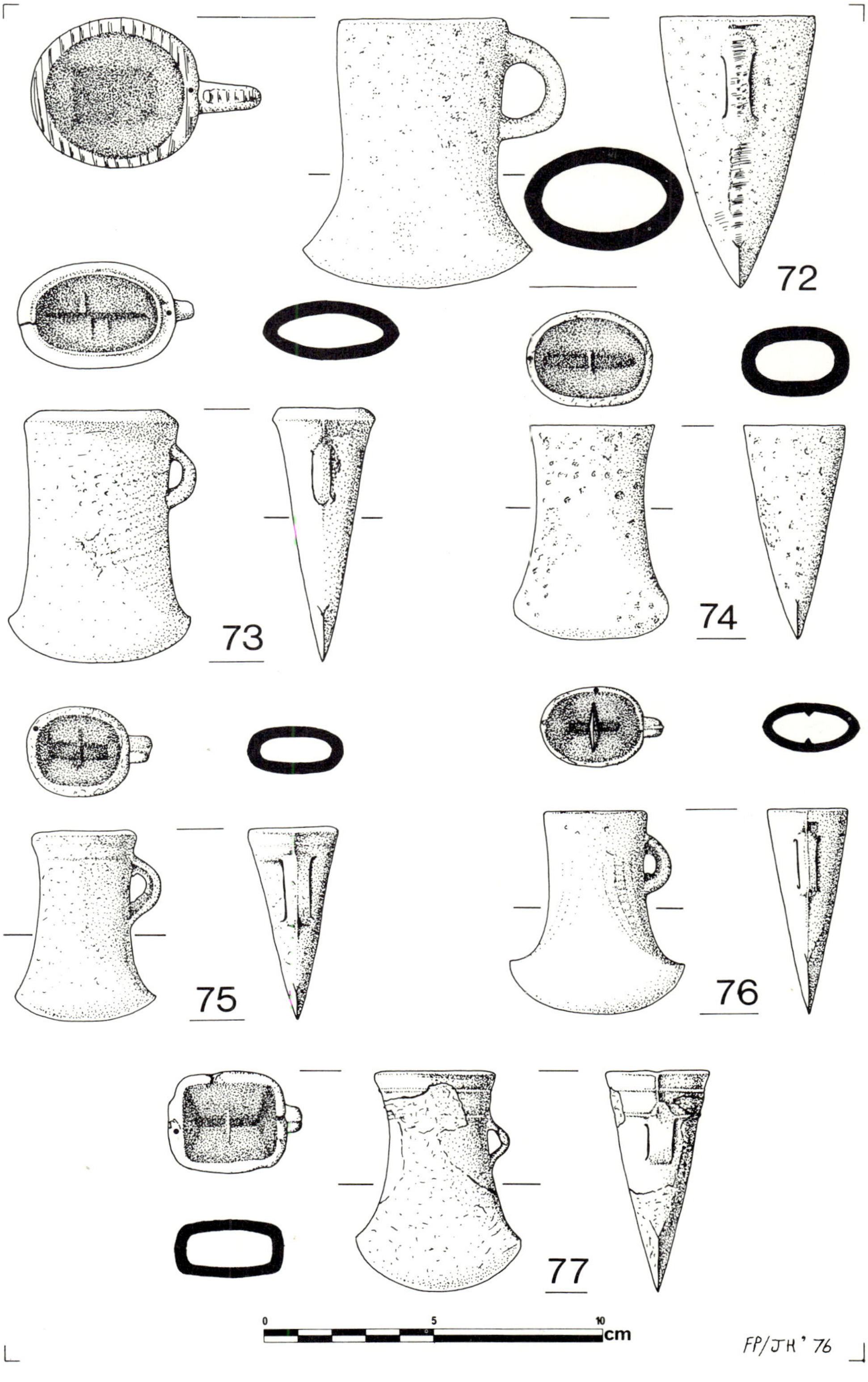

72
73
74
75
76
77
0
5
10
cm
FP/JH '76

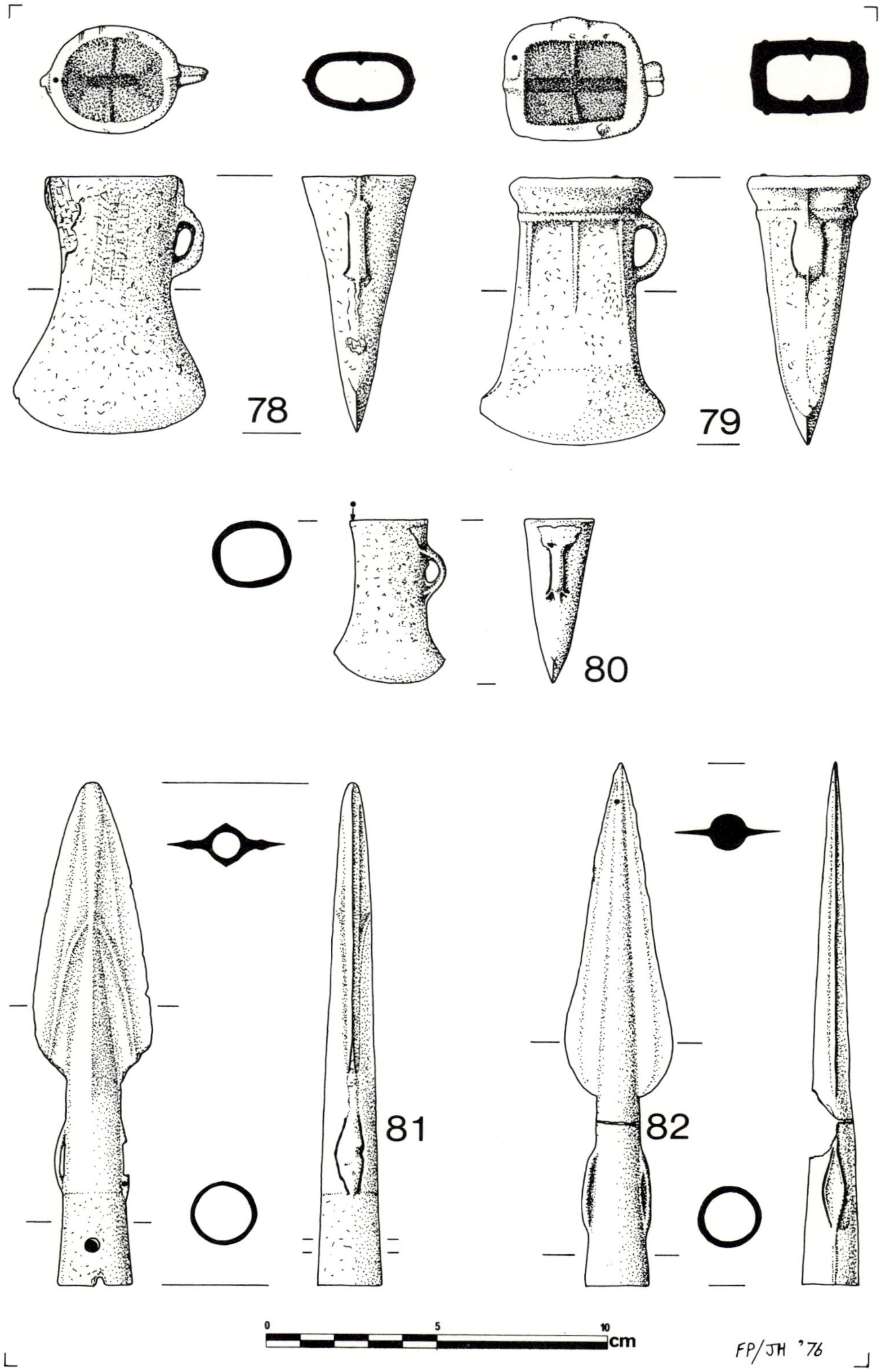
78
79
80
81
82
0
5
10
cm
FP/JH '76

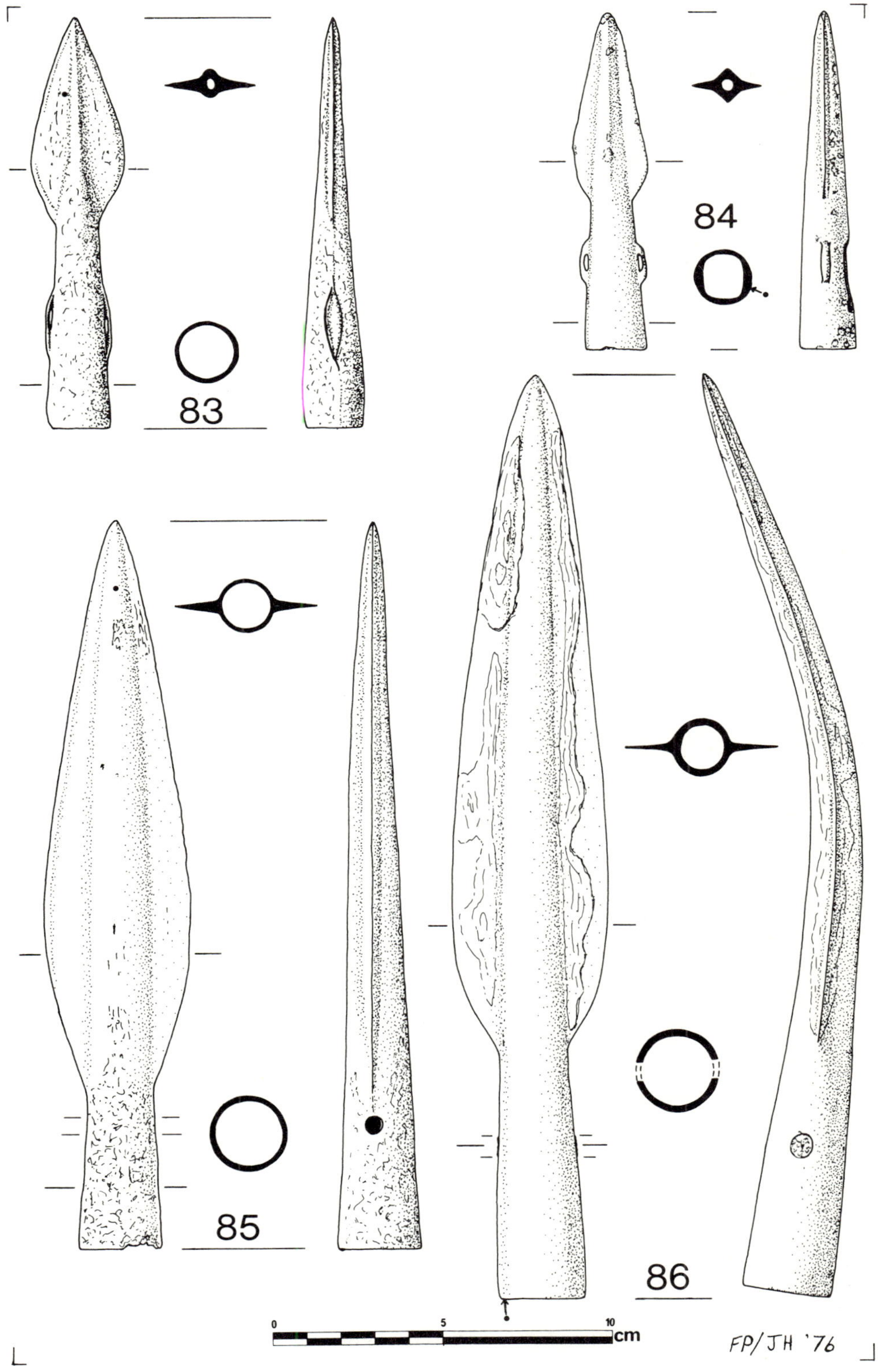
83
84
85
86
0
5
10
cm
FP/JH '76

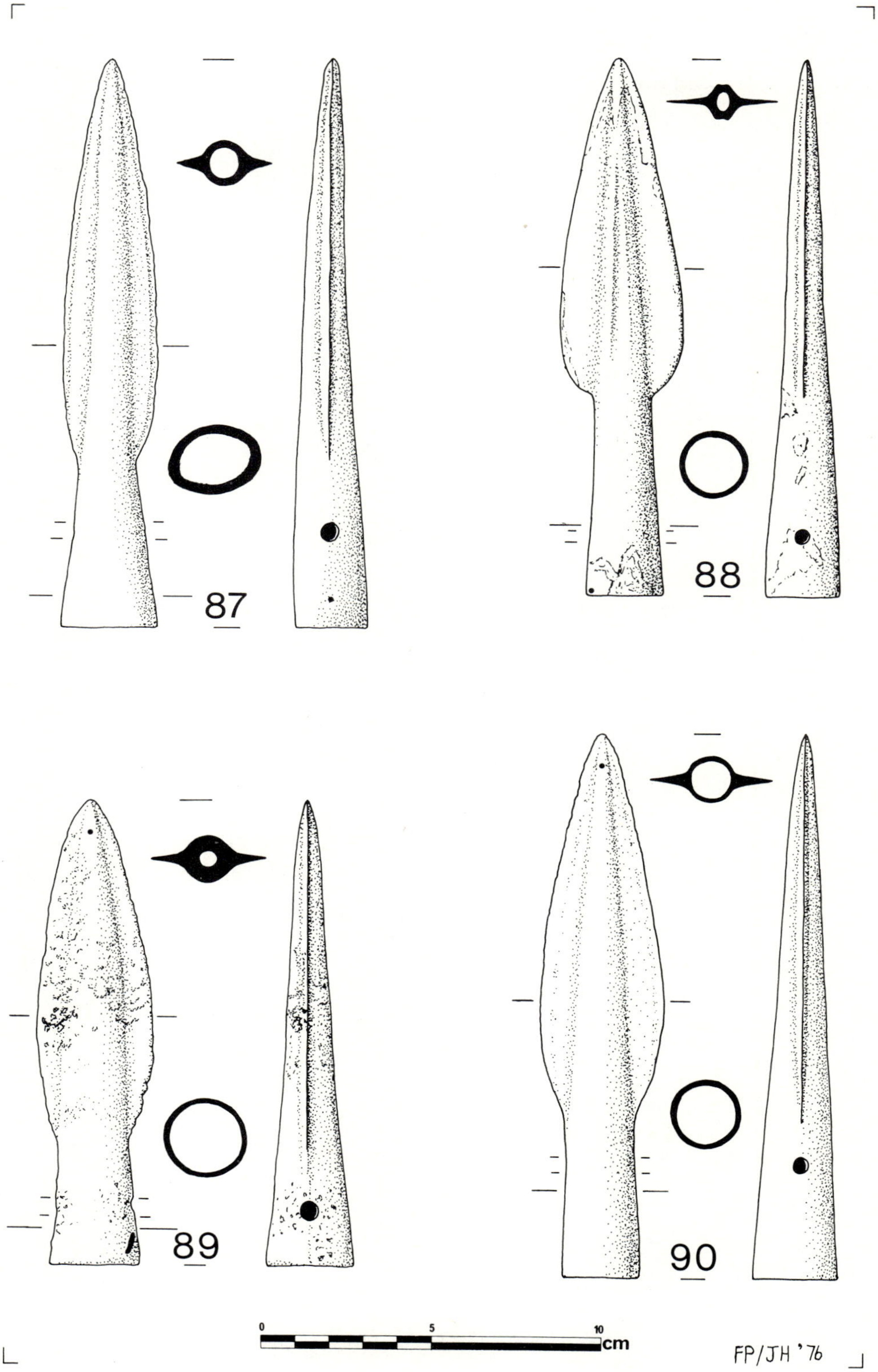

87
88
89
90
0
5
10
cm
FP/JH '76

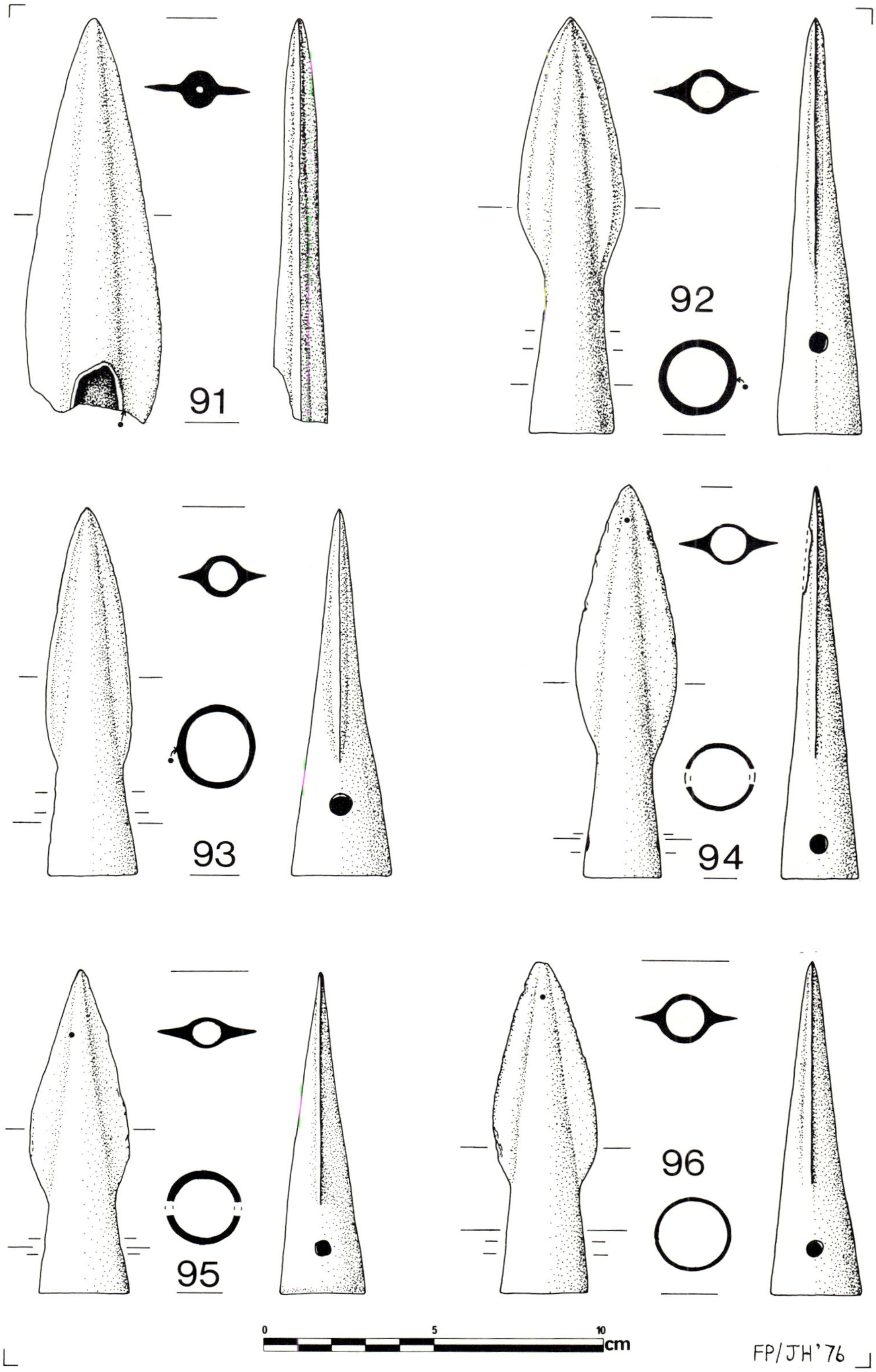
91
92
93
94
95
96
0
5
10
cm
FP/JH'76

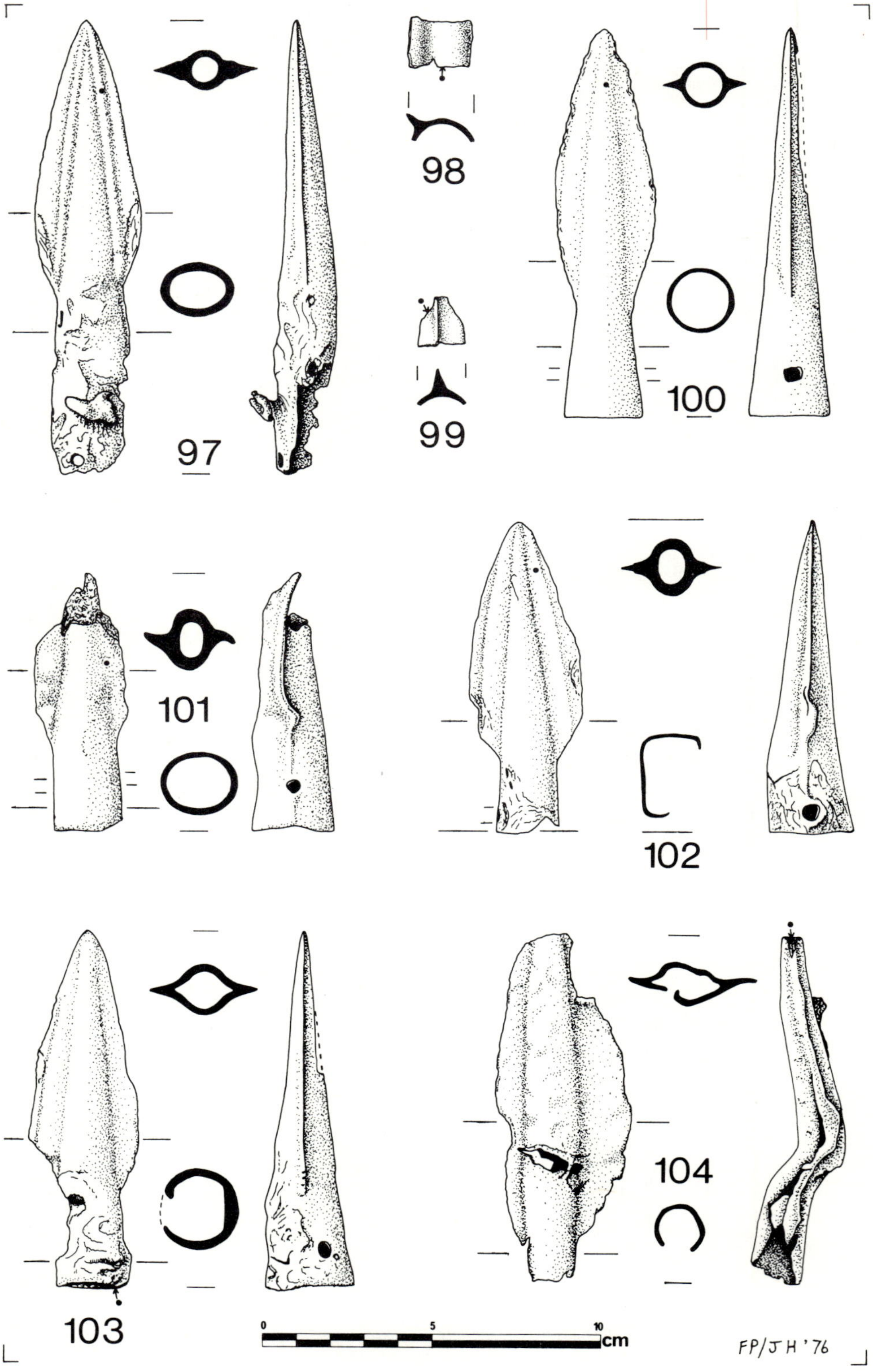
97
98
99
100
101
102
103
104
0
5
10
cm
FP/JH '76

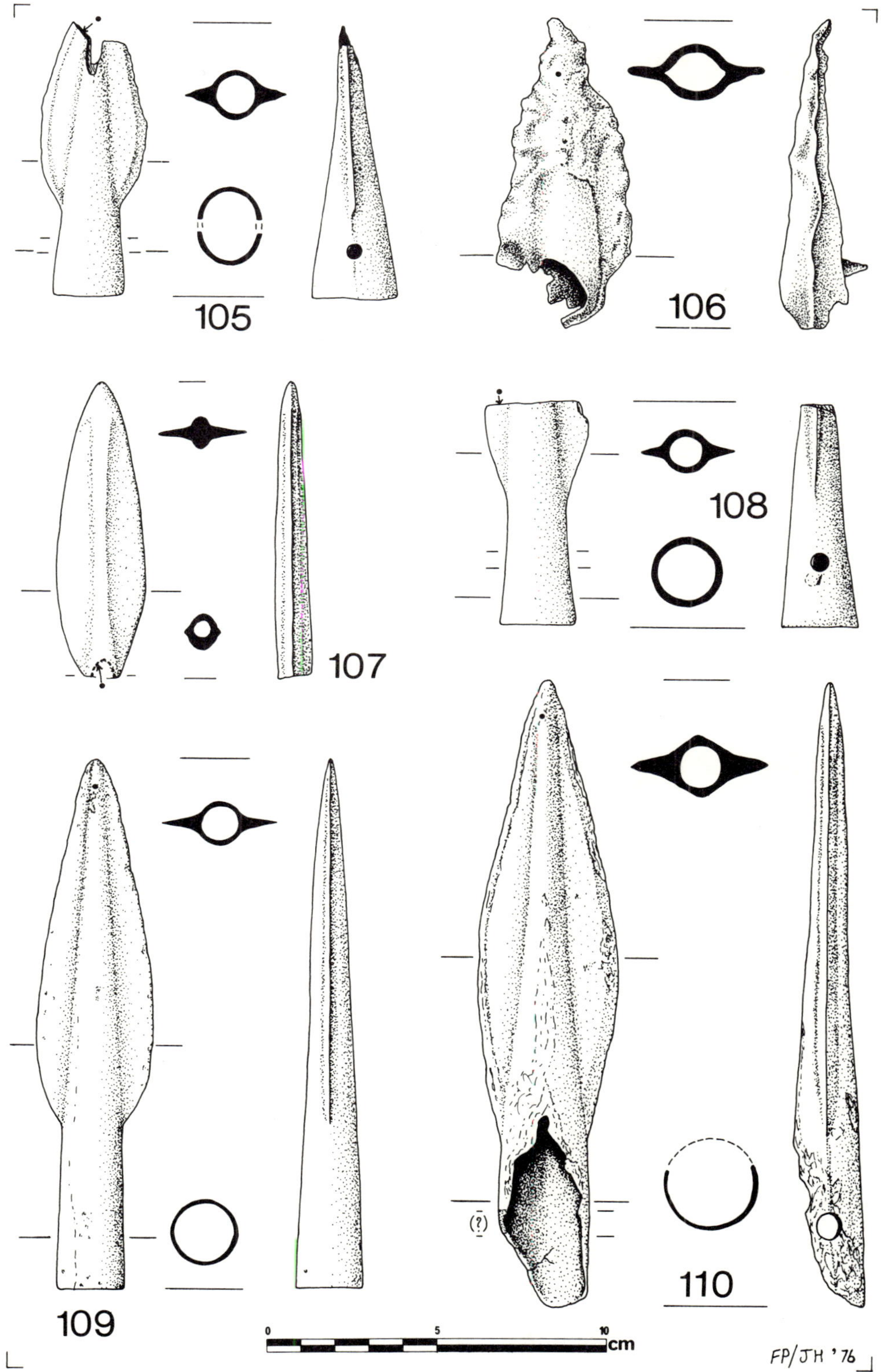
105
106
107
108
109
110
0
5
10
cm
FP/JH '76

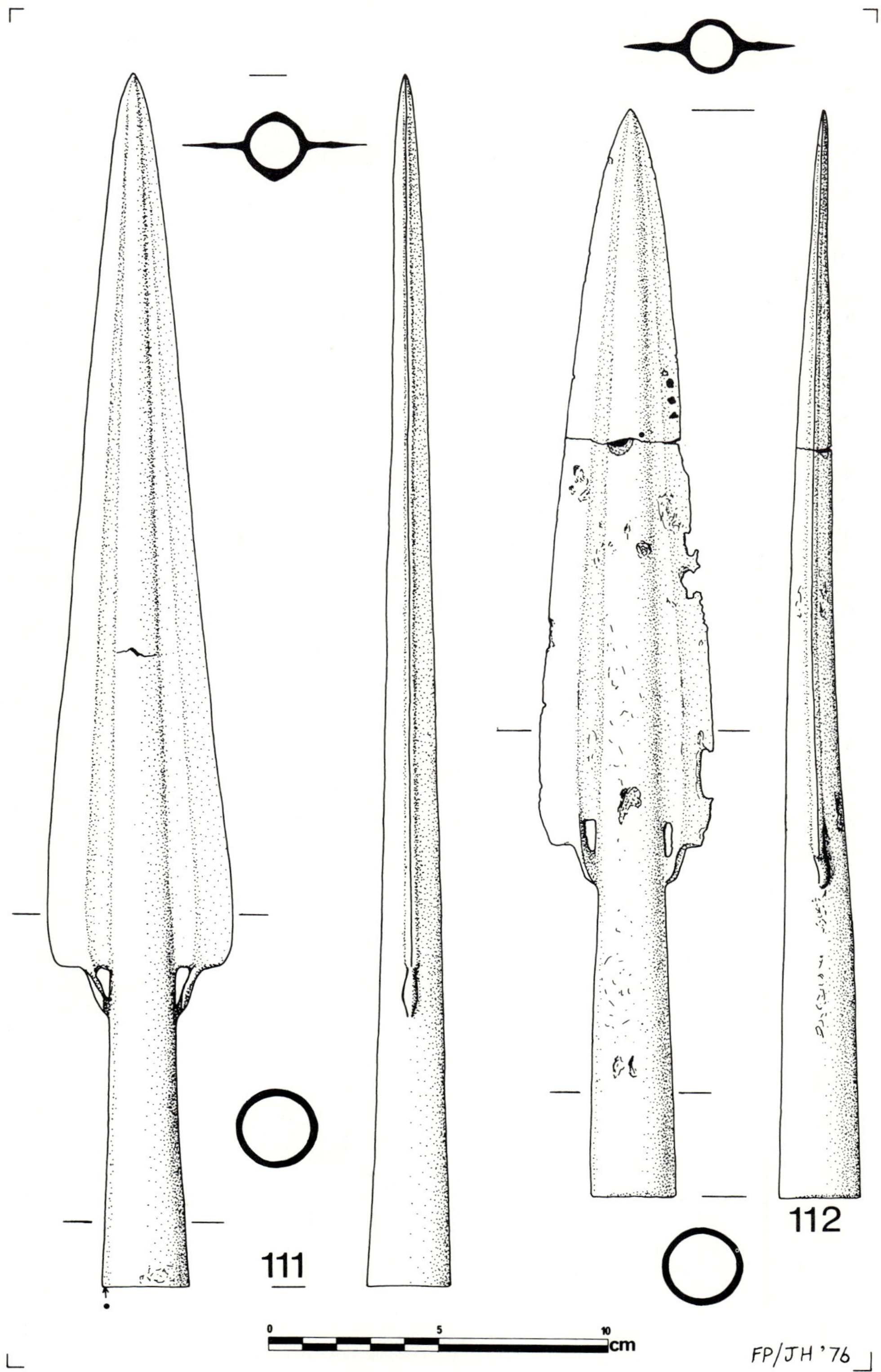
111
112
0
5
10
cm
FP/JH '76

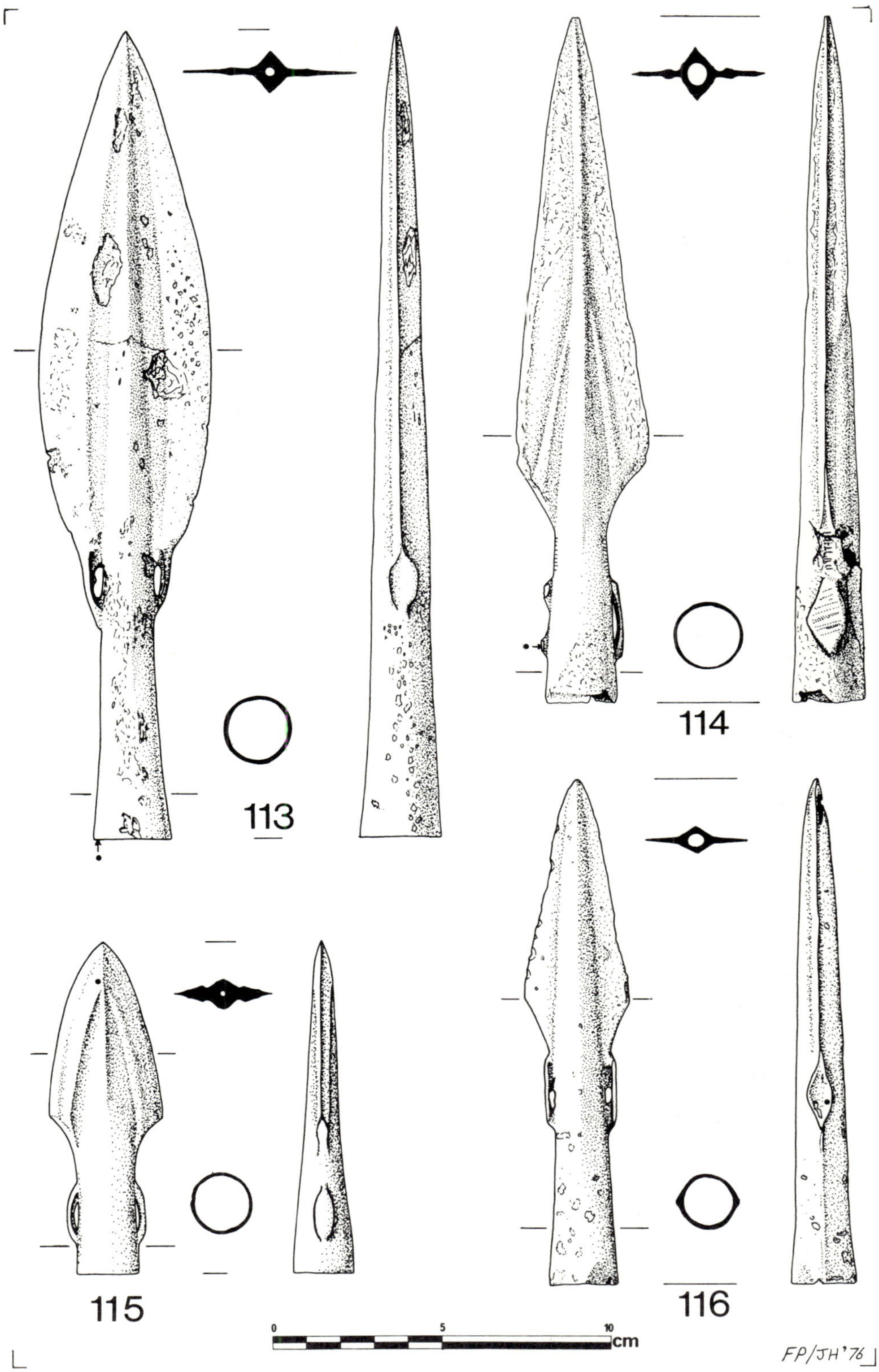
113
114
115
116
0
5
10
cm
FP/JH'76

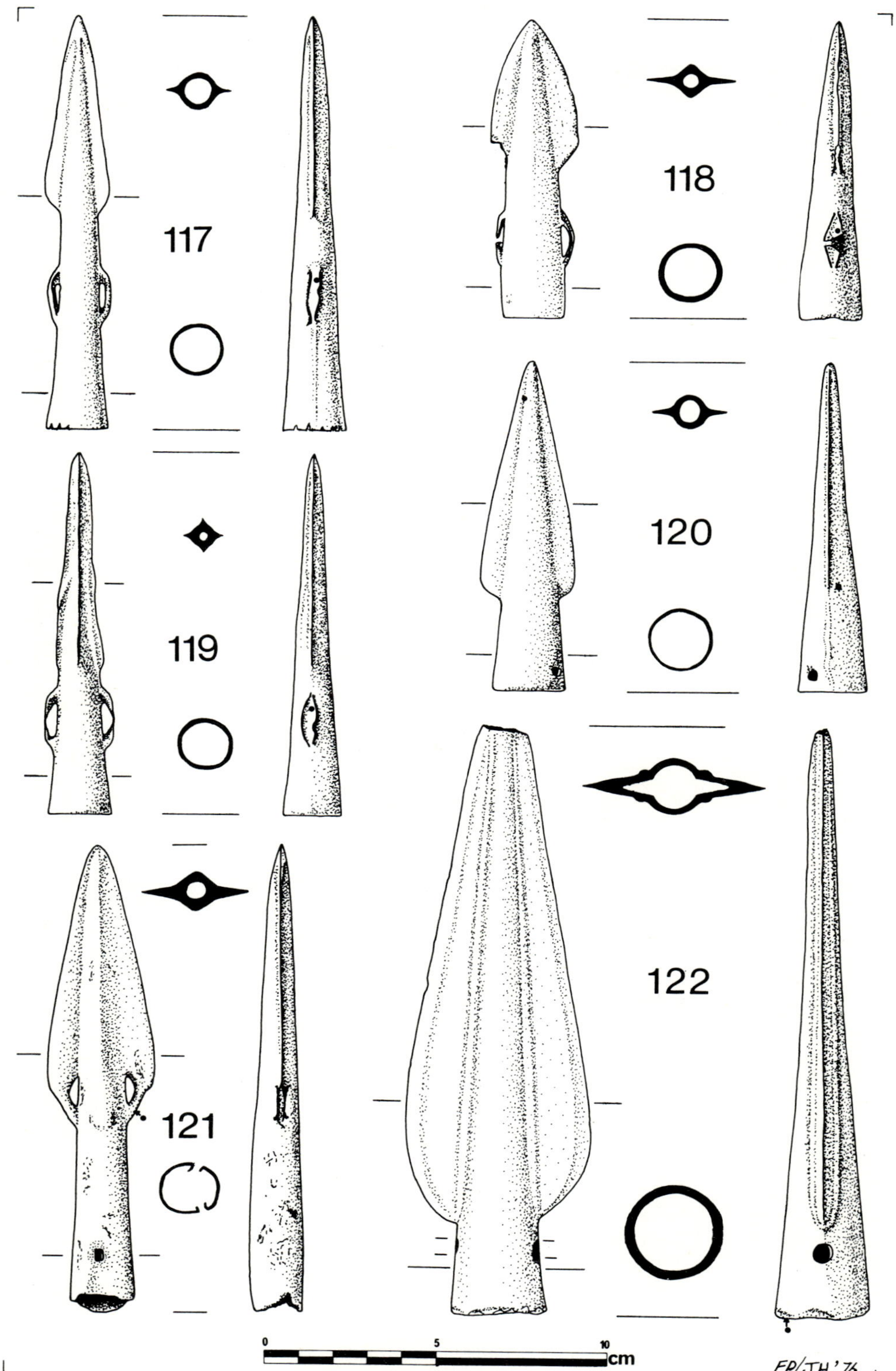
117
118
119
120
121
122
0
5
10
cm
FP/JH '76

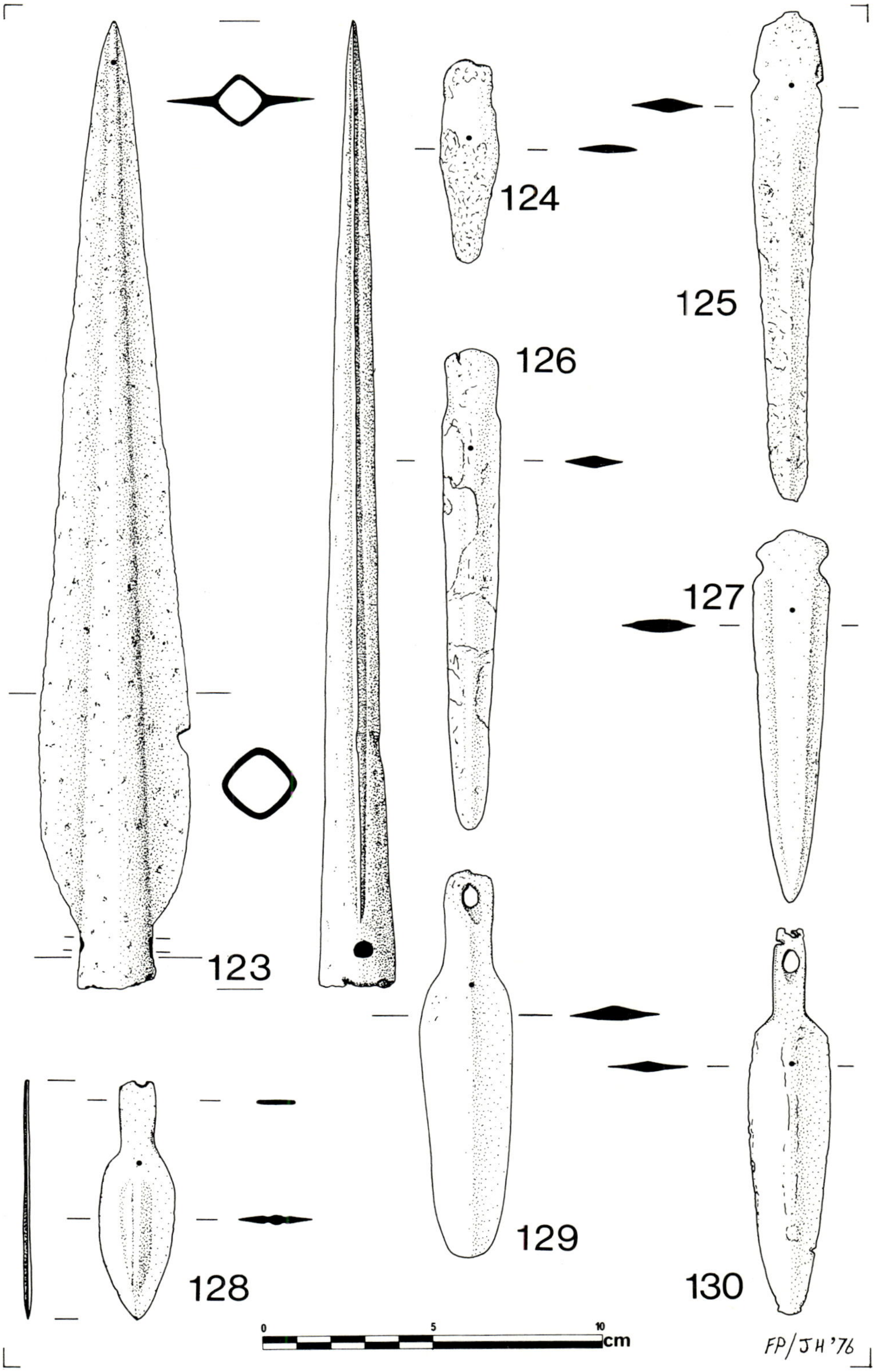
123
124
125
126
127
128
129
130
0
5
10
cm
FP/JH '76

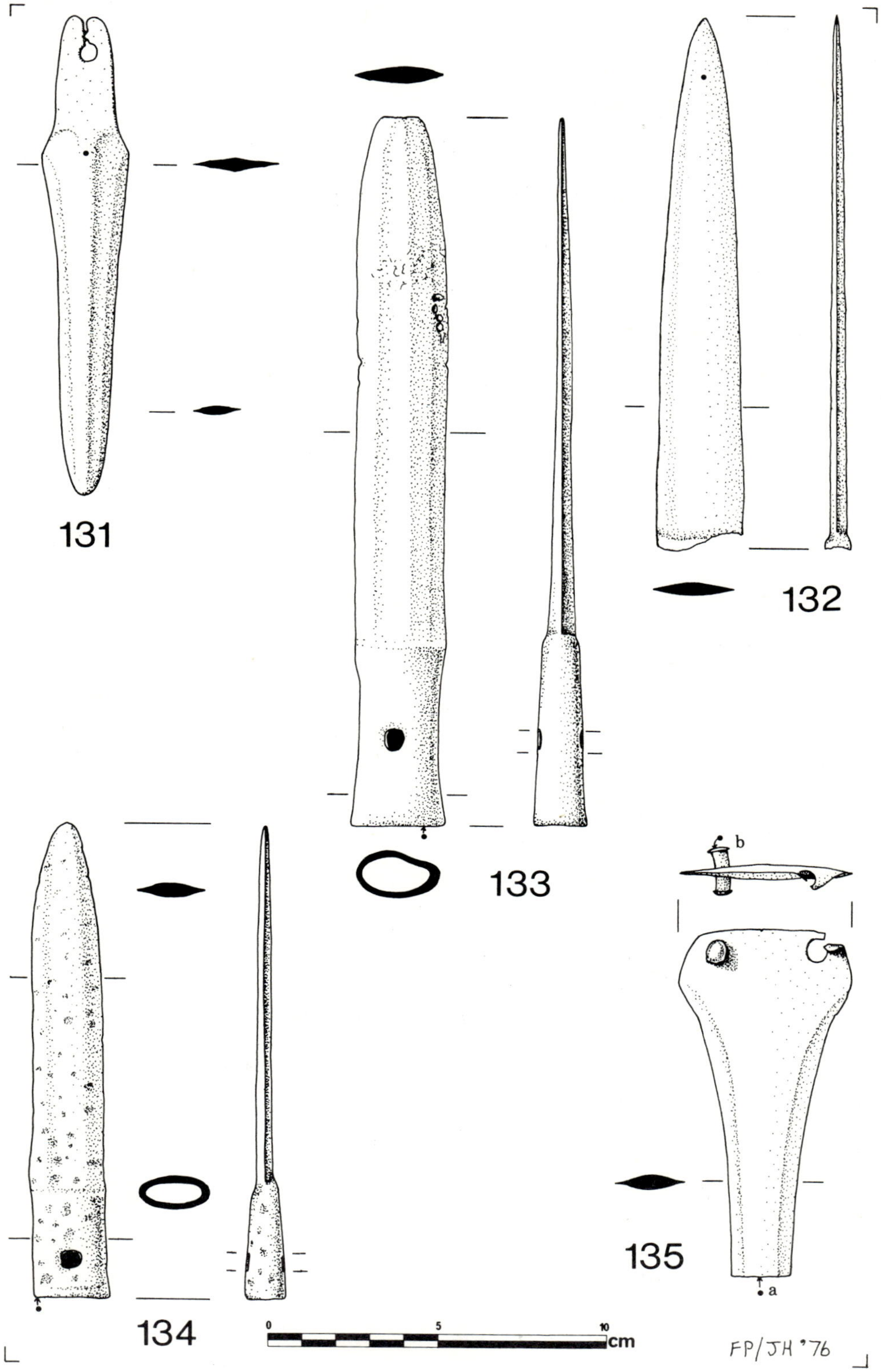
131
132
133
134
135
a
b
0
5
10
cm
FP/JH '76

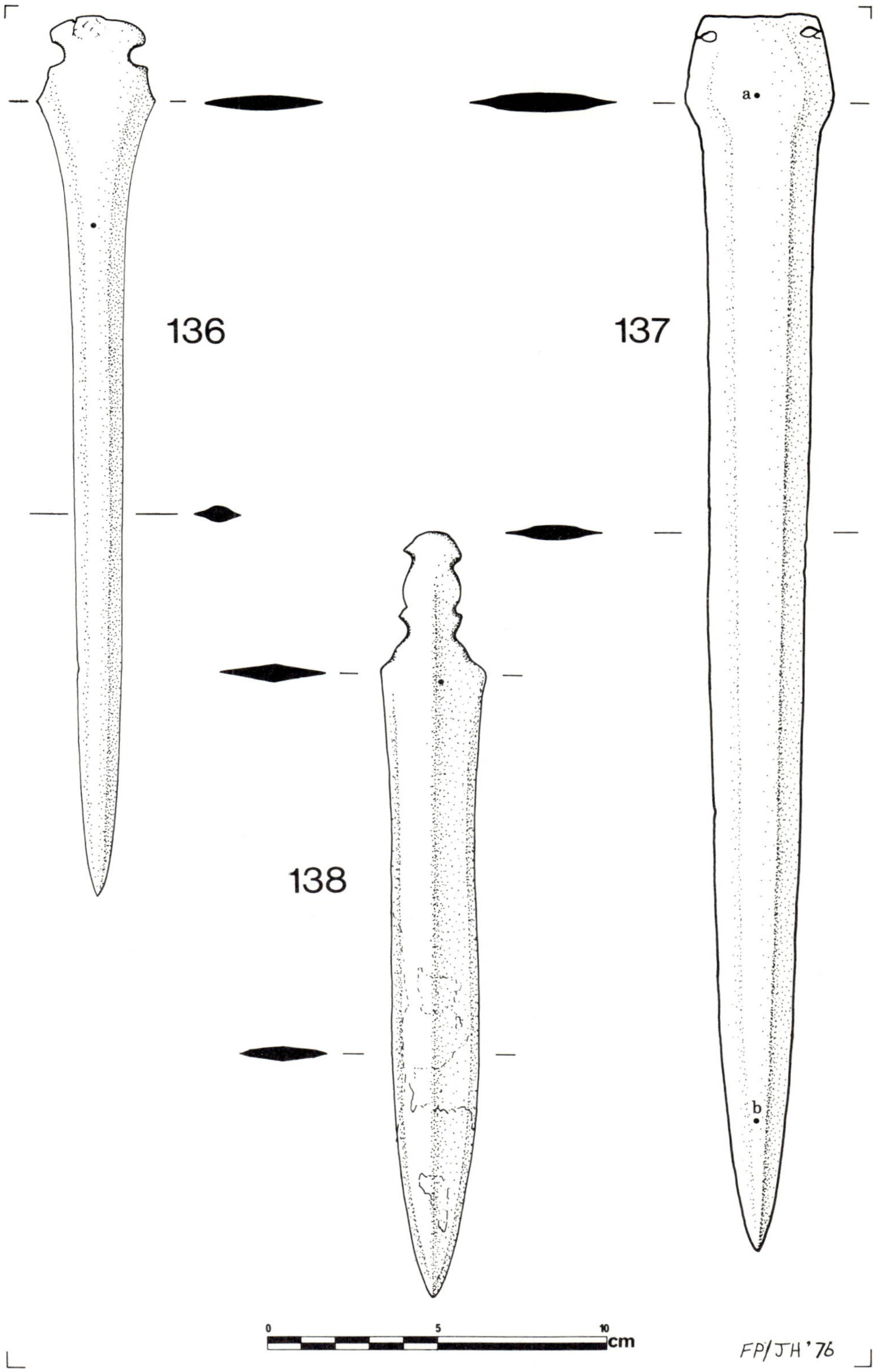

136
137
a
b
138
0
5
10
cm
FP/JH '76

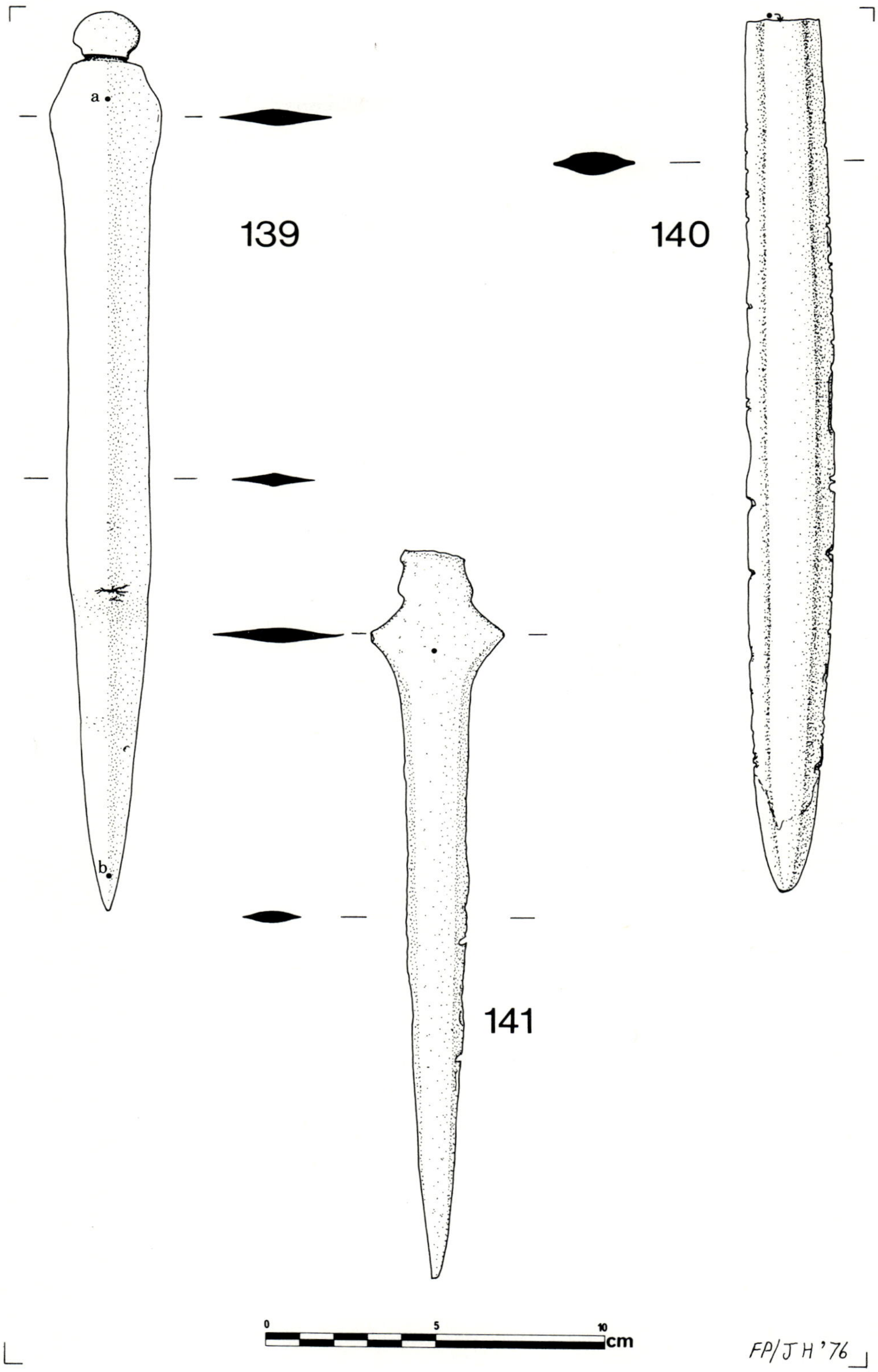
a
b
139
140
141
0
5
10
cm
FP/JH'76

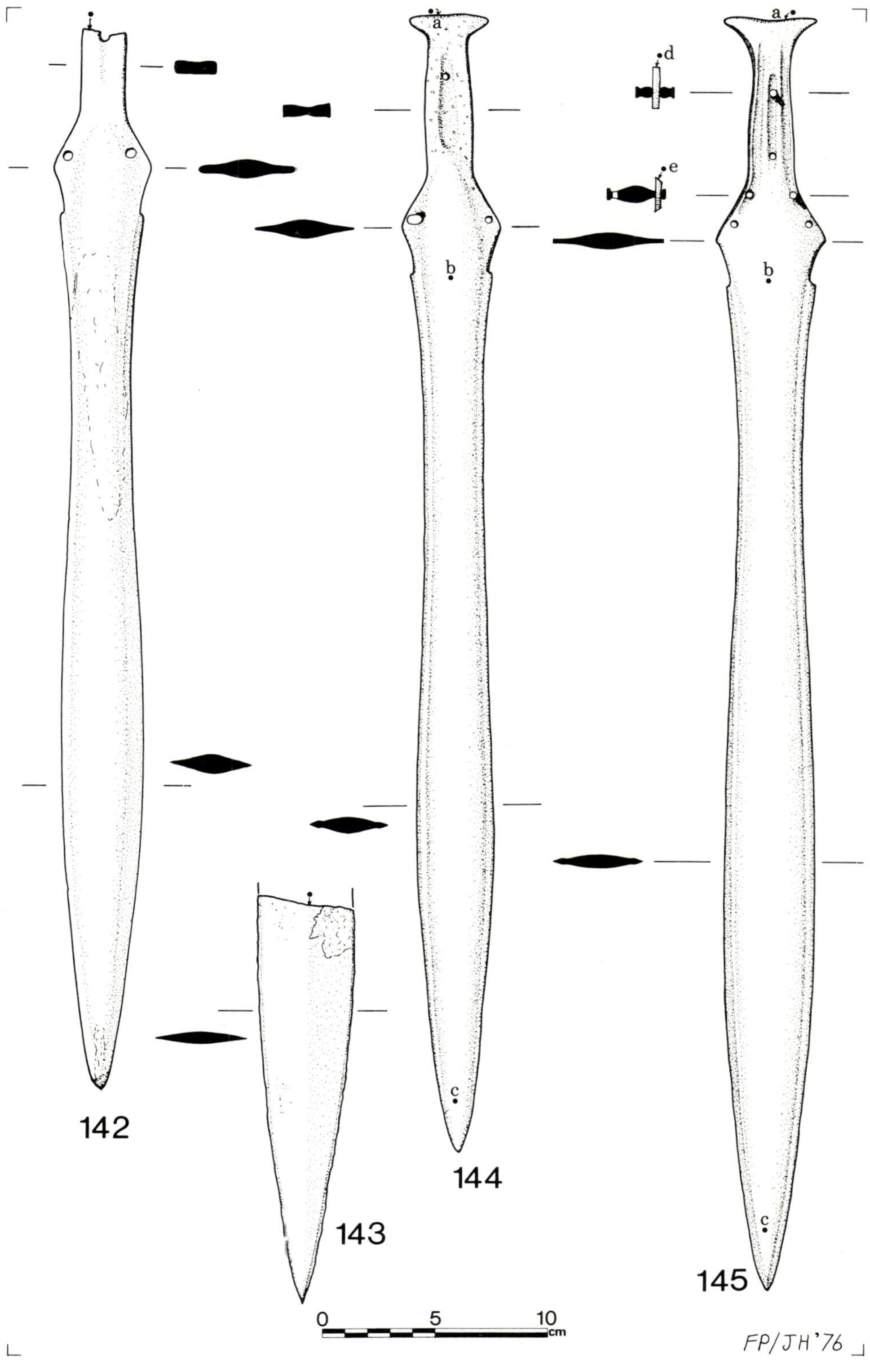

142
143
144
145
a
b
c
d
e
0
5
10
cm
FP/JH'76

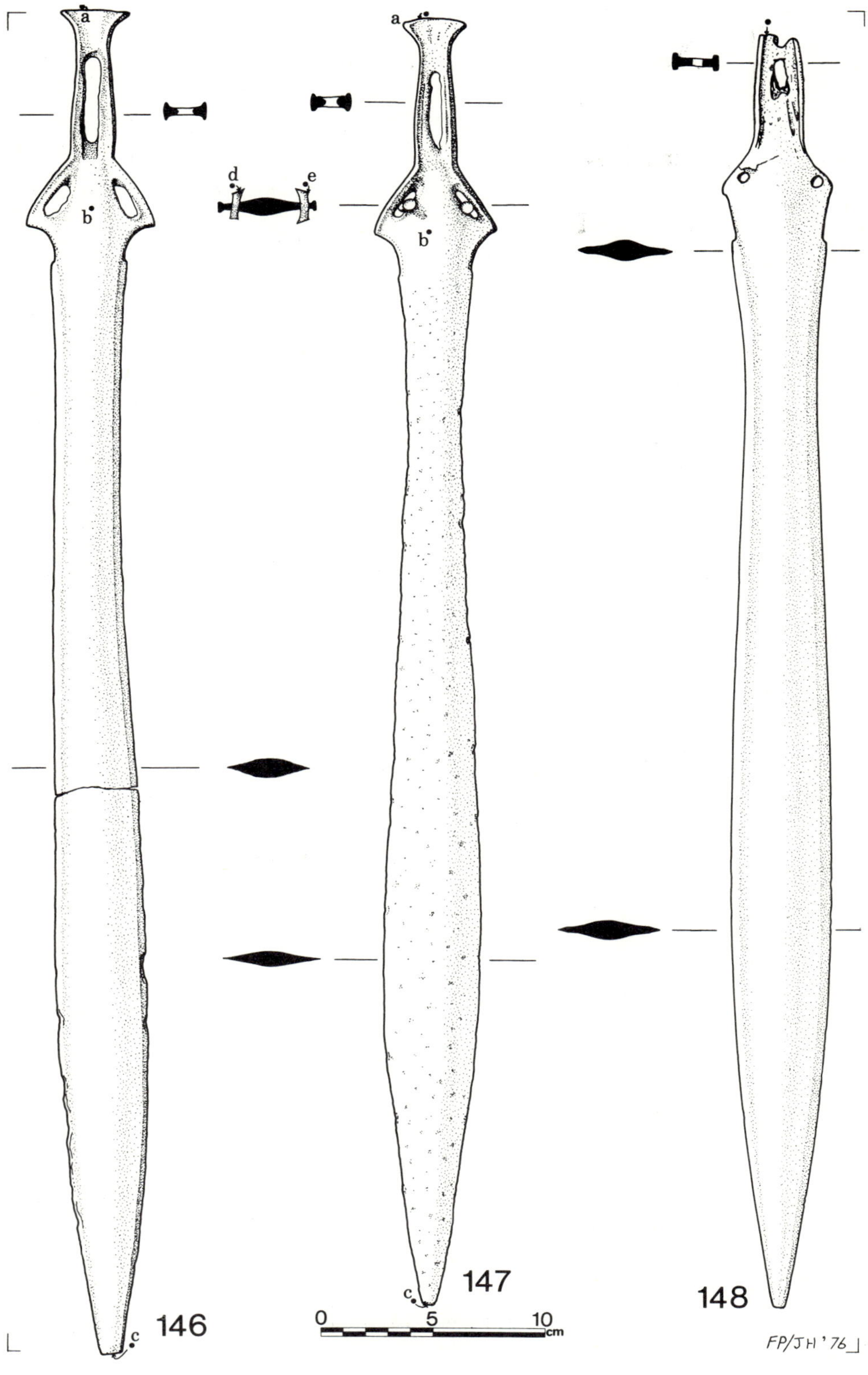

a
b
c
d
e
146
147
148
0
5
10
cm
FP/JH '76

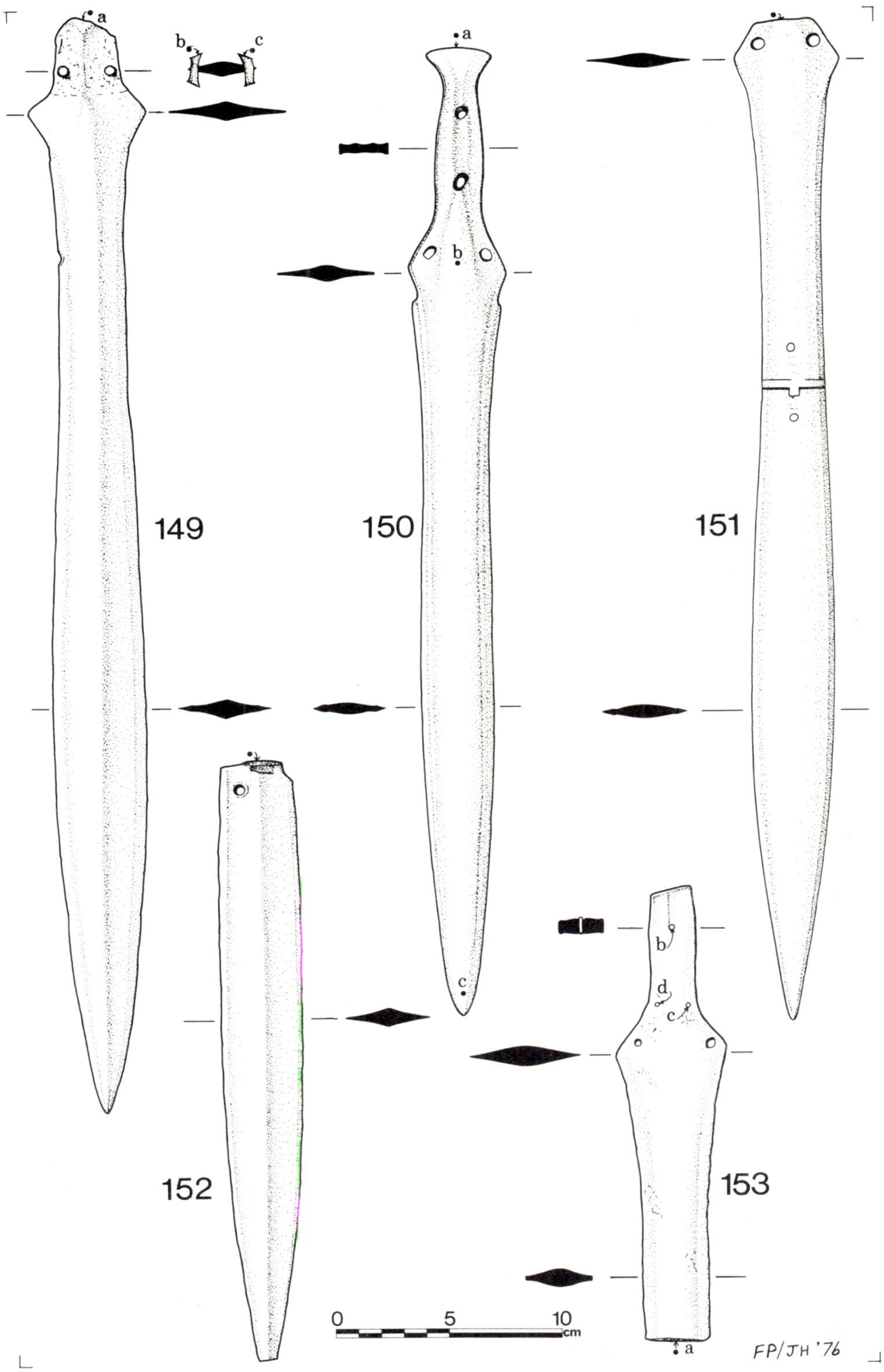
149
150
151
152
153
a
b
c
d
0
5
10
cm
FP/JH '76

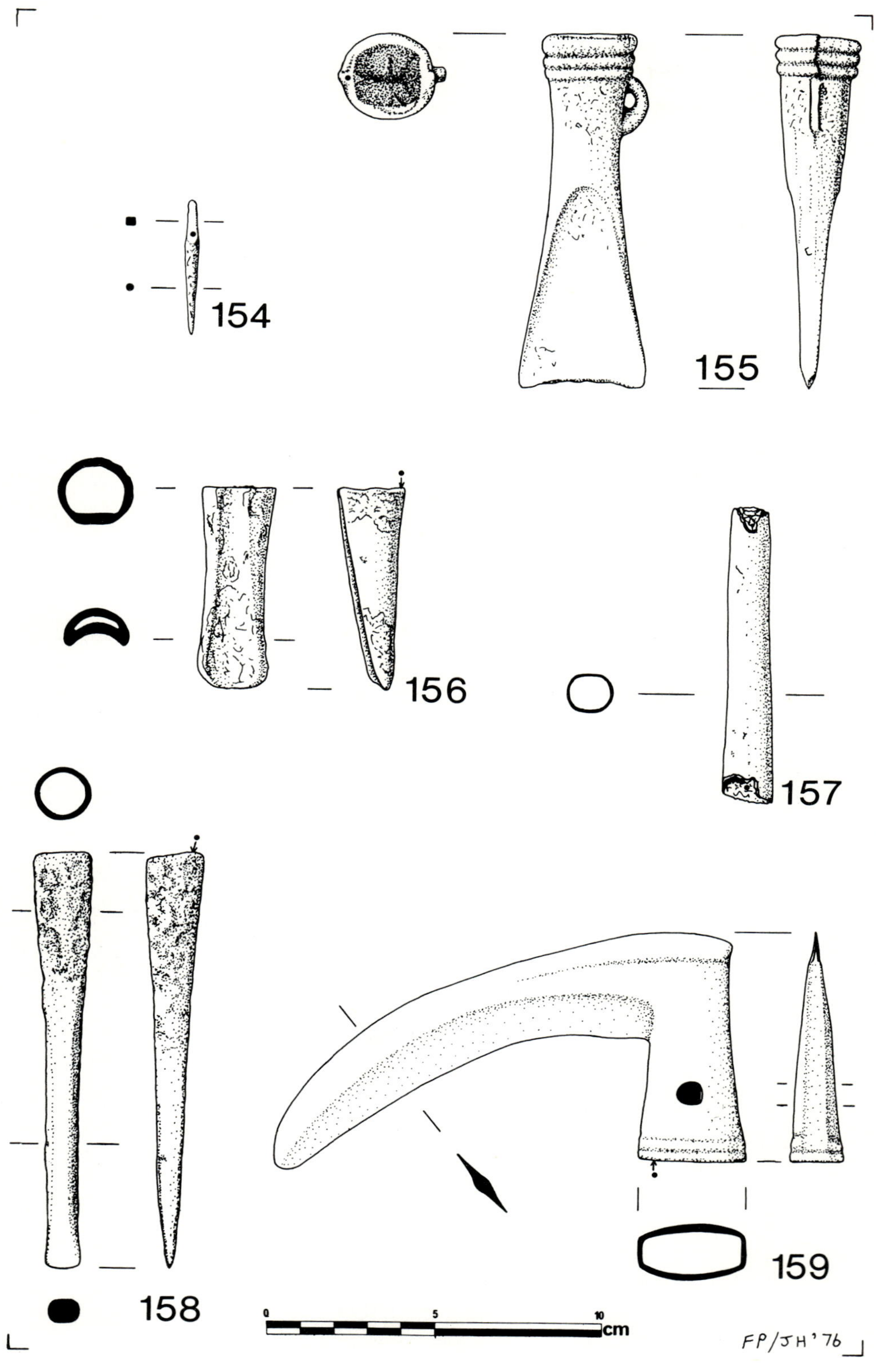
154
155
156
157
158
159
0
5
10
cm
FP/JH'76

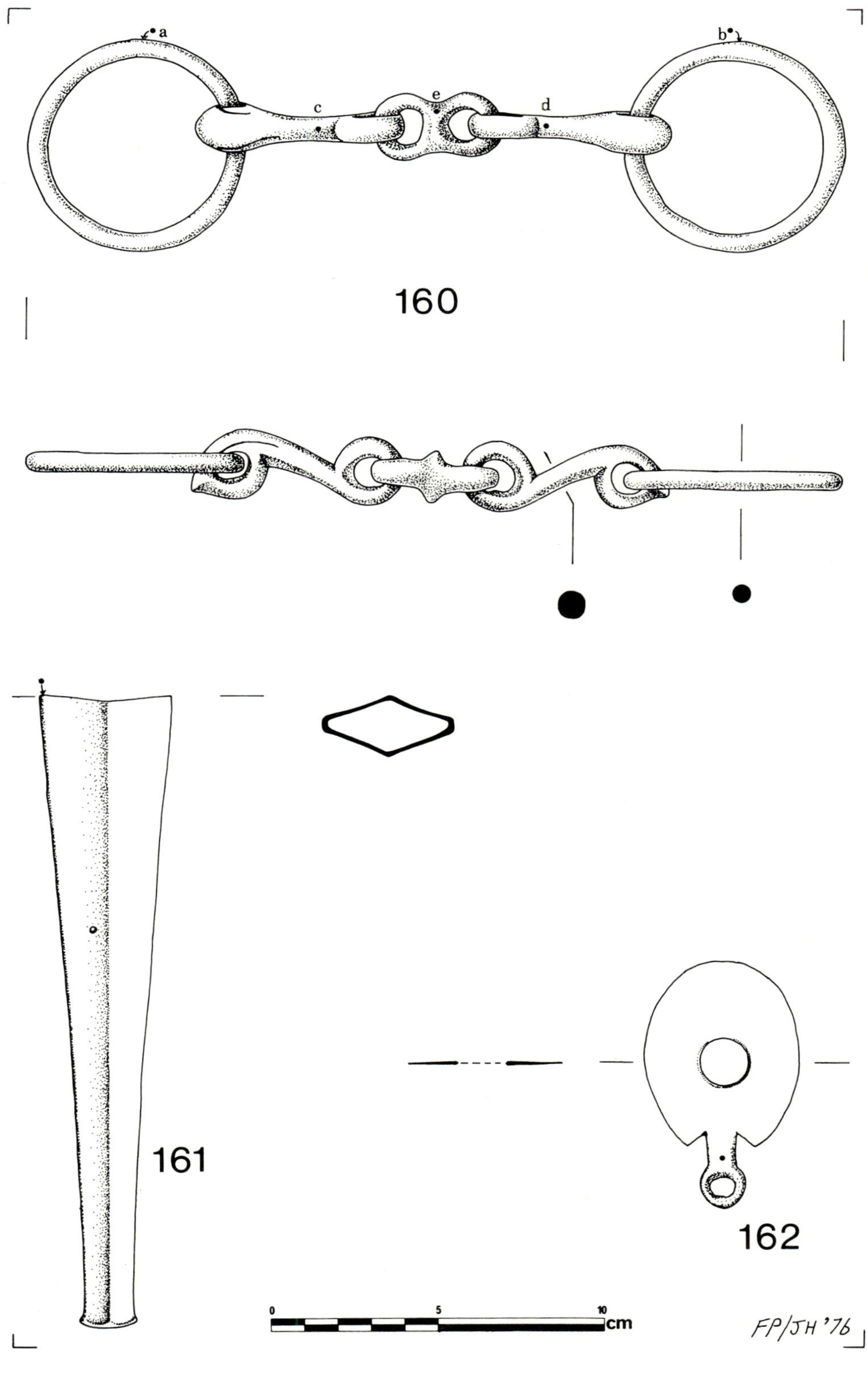

a
b
c
e
d
160
161
162
0
5
10
cm
FP/JH '76

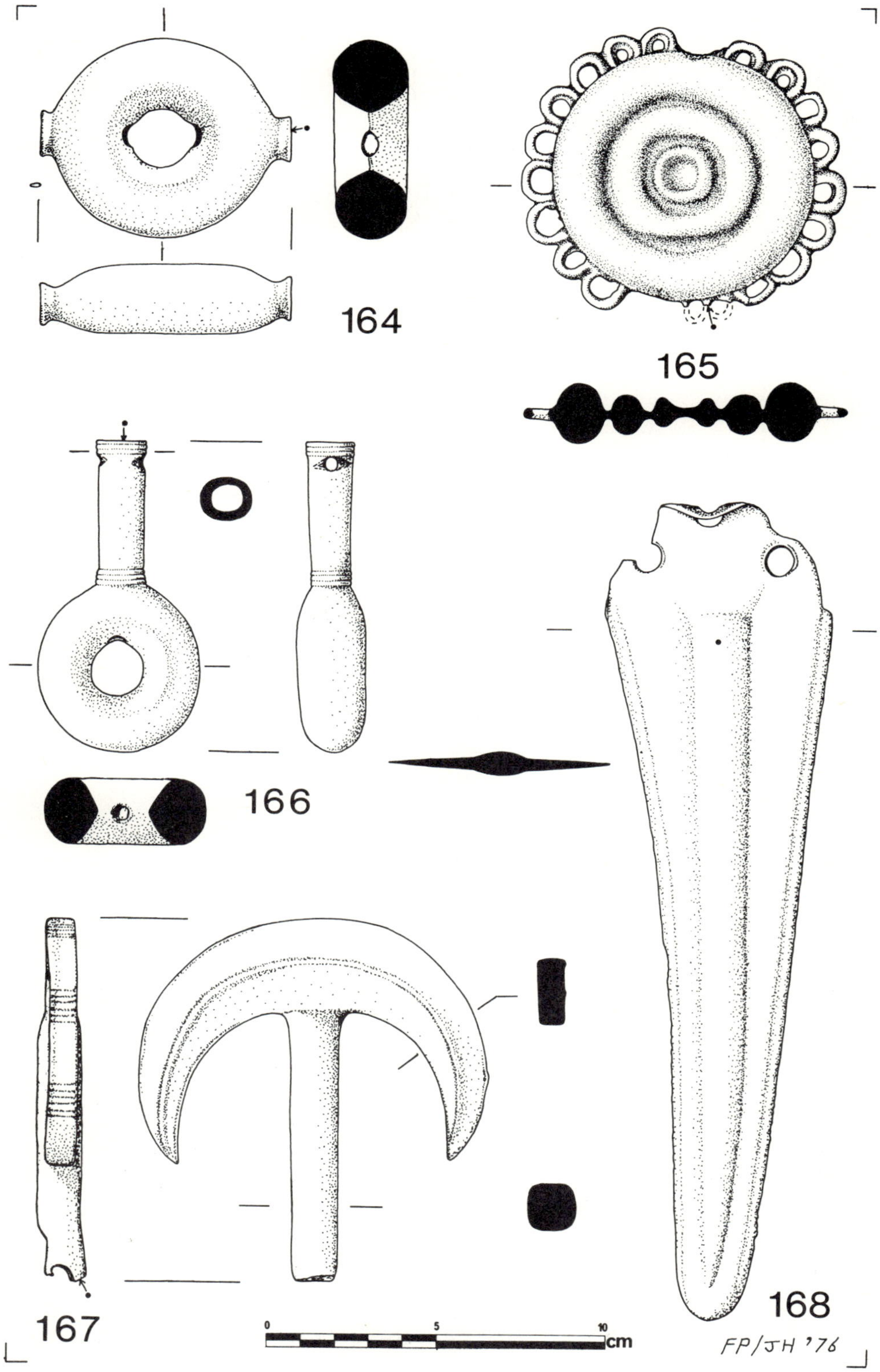
164
165
166
167
168
0
5
10
cm
FP/JH '76

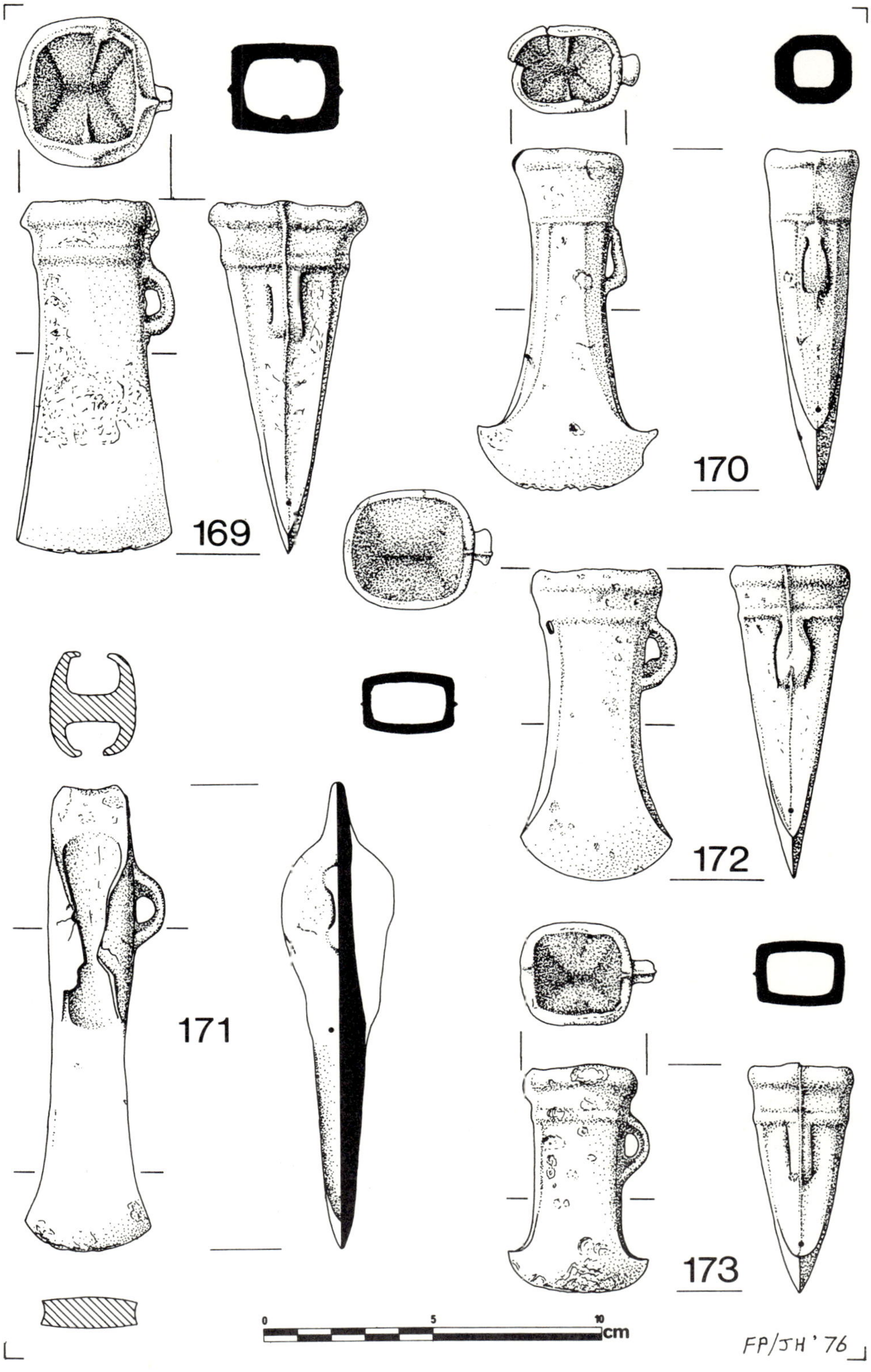
169
170
171
172
173
0
5
10
cm
FP/JH '76

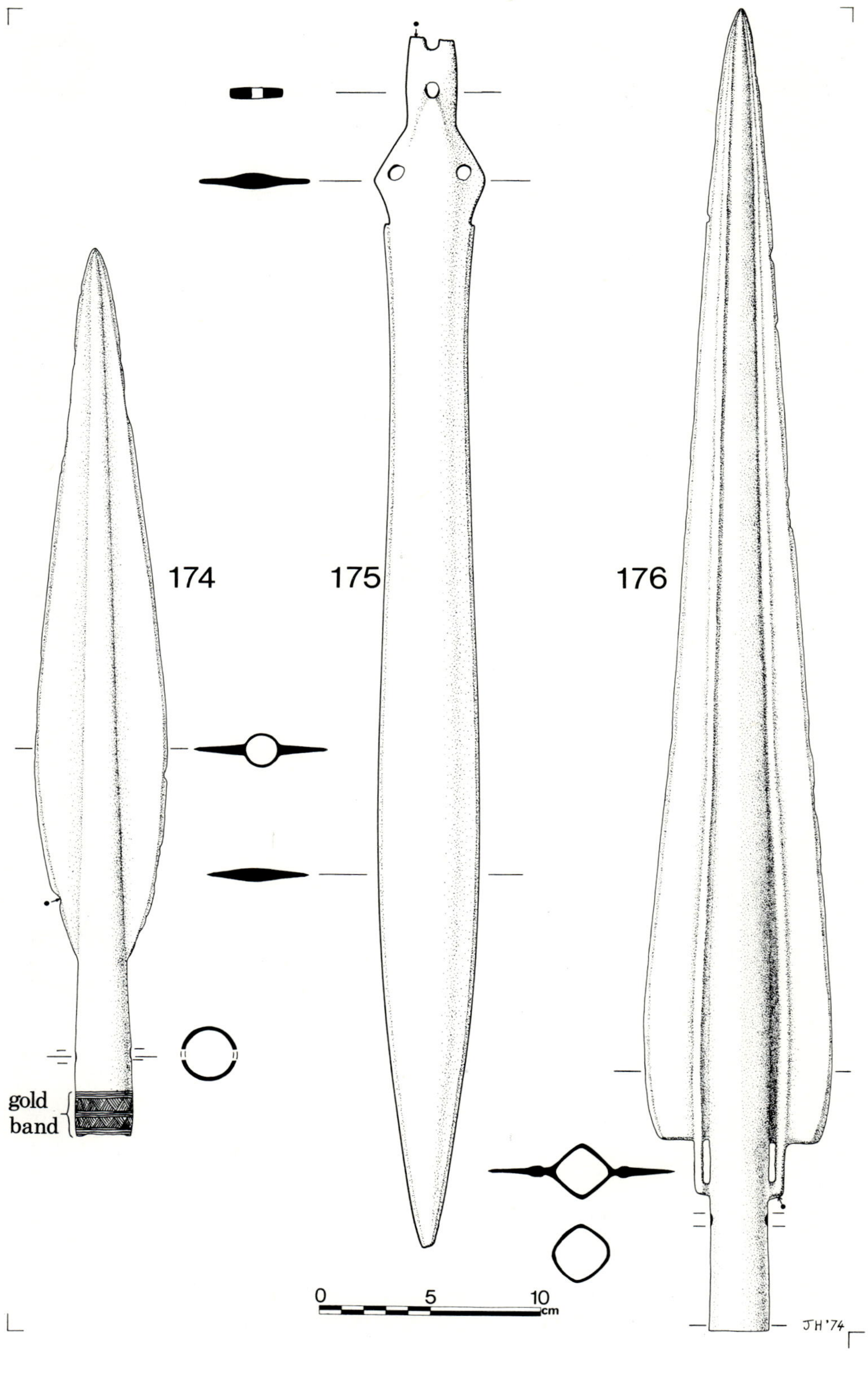
174
175
176
gold
band
0
5
10
cm
JH '74

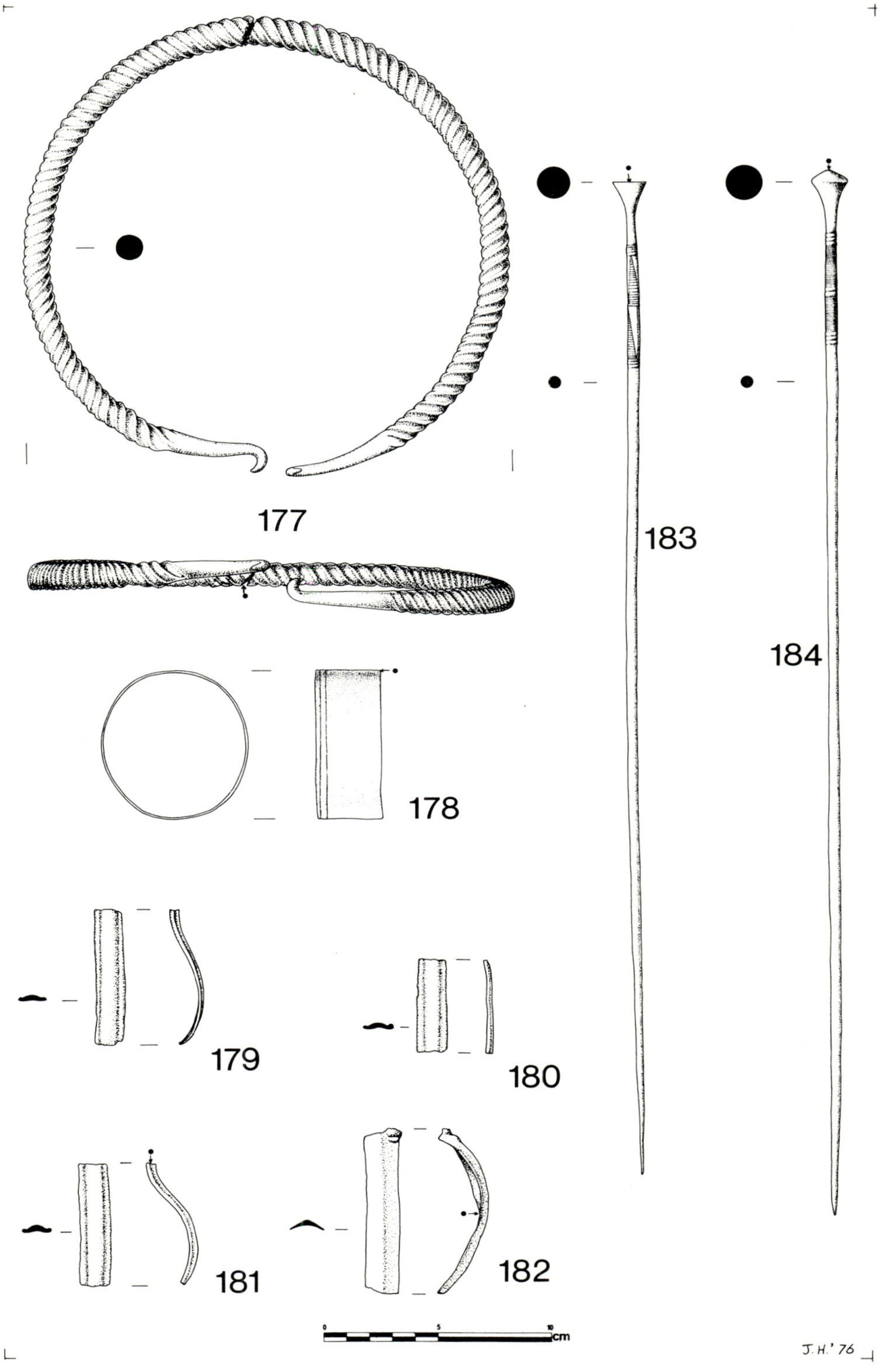
177
178
179
180
181
182
183
184
0 5 10 cm
J.H.'76

The Composition of the Copper Alloys

Dr. Paul Craddock

The prehistorian wishing to learn about the early metallurgy of Britain is well served with information both in the excellent general survey of Tylecote (1962) and in a host of more specialized works of which the analyses of Middle and Late Bronze Age metalwork by Brown and Blin-Stoyle (1959) are the most pertinent to this study. With such a body of knowledge in existence, the role of analyses in a publication such as this is more to aid the description and understanding of the objects themselves than to create new theories on ancient metallurgy, or to extend knowledge into completely new fields. The material analysed for this catalogue does not include any major types that were previously unanalysed, and with the exception of the antimony bronzes discussed below, no new alloy types have been encountered. Thus, few general conclusions on Bronze Age metallurgy will be given in this report. These analyses, however, do form a very valuable part of a continuing series of analyses of Bronze Age metalwork from Britain and Western Europe which have been performed by the British Museum Research Laboratory. It is hoped to publish the results reported here, together with other Bronze Age analyses, in a more general paper to be included in the *Proceedings of Atlantic Colloquium V*—a report on a symposium held in Dublin in 1978 to discuss the origins of metallurgy in Atlantic Europe.

The fact that the ROM bronzes have been analysed by the same methodology and personnel as the other, comparable, Bronze Age metalwork may seem of little consequence; but unfortunately recent inter-laboratory comparisons on standard bronzes have revealed great discrepancies, not only in the elements searched for and detected, but also in the actual compositions determined (Chase 1976). These discrepancies were noticeably greatest for the trace element concentrations which are usually relied upon when the grouping or provenancing of bronze artifacts is attempted by prehistorians. Therefore to formulate theories on the basis of data that has been derived from a number of analytical sources is potentially very dangerous. Thus, by including the ROM's important collection in metal being studied in Britain itself, we have ensured that the analytical groupings will relate to the Bronze Age and not to different laboratories.

Methods of Sampling and Analysis

The samples were drilled where possible from the main body of the artifact using a portable jeweller's drill with a size 60 bit. This method gives a 1 mm diameter hole, the position of which is marked on the illustrations by an outsize dot. In order to avoid the surface and any corroded material, the sample was not collected until bright turnings of metal were observed. Since the composition of corroded metal need bear little relation to that of the original, analysis of corroded metal can give very misleading results, and is best not attempted (Craddock 1976). Metal alloys are rarely homogeneous throughout, this fact being especially true of bronze, and thus great care must be taken in the selection of sampling positions to ensure that the composition of the sample relates to the composition of the object as a whole. The principal forms of segregation are discussed by Bailey (1966), but may be summarized here. For leaded tin bronze they include: (1) gravity segregation—where the heaviest component of an alloy, the lead, sinks to the bottom of the mould and the lightest, the tin-rich alloy, tends to rise; (2) normal segregation—where the tin-rich part of the melt tends to solidify last, as it has the lowest melting point, and is therefore concentrated in the top and centre of the casting, away from the sides; (3) inverse segregation—a phenomenon also known as "sweating" whereby the tin and lead can become concentrated on the surface of the metal. In taking samples it is, therefore, clearly important to avoid areas likely to be affected by segregation and generally the desired results can be achieved by sampling away from surfaces and extremities of castings. Where possible, therefore, the drilling was taken near the centre of the casting and, typically, was drilled to a depth of about 5 mm, thus ensuring that an adequate cross-section of the metal had been sampled at the drilling point. Usually about 20 mg of metal were removed at each drilling.

As segregation is dependent not only on composition size but on casting conditions, its presence or absence can give information on those conditions. Thus, several of the swords and one of the axes, believed to be heavily leaded, were sampled in several places along their length (including their extremities) to obtain information on casting conditions.

The samples were prepared and analysed by atomic absorption spectrometry using standard apparatus and methods. The exact procedure followed has been recently described in detail (Hughes et al. 1976) and will not be repeated here. The arsenic and bismuth content was determined using a graphite furnace attachment which gave greater sensitivity for these elements; otherwise the usual flame methods were adopted.

The precision of the analyses is ±1 per cent for the major elements and ±20 per cent for the trace elements. The sensitivity of the method enabled the elements to be detected down to at least 0.005 per cent in the metal.

Results and Discussion

The earliest ROM material is represented by the broad-butted axes, only one of which is of tin bronze, the remainder being of copper or of copper with some arsenic. This range of alloy types is similar to that reported for broad-butted axes by Britton (1963). The halberd (no. 168) is also of arsenical copper. In each case the arsenic is associated with a relatively high silver, antimony, and nickel content, a condition which suggests the use of fahlerz ores of secondary enrichment in which high concentrations of these elements are common (Tylecote 1962). Although the arsenic levels are low compared with some other European arsenical coppers (Craddock 1977), they are significantly higher than in the remainder of British Early Bronze Age material. These levels strongly suggest that copper ores were being carefully selected at this early period to include material which when smelted would give a copper rich in arsenic. Arsenic, like tin, considerably helps the casting and working properties of the metal.

In Britain, however, tin very quickly replaced arsenic as the chief alloying metal used with the copper. The tin content of the Early Bronze Age ROM material varies from 8.6 to 14.7 per cent with an average content of 11.5 per cent (Fig.1).

The tin content of the Middle Bronze Age ROM material varies from 7.1 to 18.2 per cent with an average content of 11.6 per cent (Fig. 2). It is interesting to note that the axes have significantly more tin (average 12.7%) than the other types of metalwork (average 9.3%) (Fig. 2).

The tin content of the Late Bronze Age ROM material (excluding that alloyed with antimony or zinc) varies from 2.9 to 16.4 per cent with an average of 9.4 per cent (Fig. 3). Most of the Late Bronze Age ROM material contains more than 1 per cent of lead (Fig. 4).

The tin histograms (Fig. 1, 2, and 3) show a reasonably normal distribution of concentrations around a mean, whereas the lead histogram (Fig. 4) indicates that there is no preferred concentration, but instead a gradual fall off in the number of bronzes with from 0 to 10 per cent lead content, and then an essentially random distribution up to 27 per cent; these percentages suggest that no attempt was being made to achieve a particular lead content. The patterns of tin and lead concentrations are very similar both to those for contemporary Continental Late Bronze Age metal from Spain, and those for Geometric bronzes from Greece (Craddock 1976 and 1977).

Three Late Bronze Age artifacts (nos. 62, 155, and 161) are all of a bronze alloy in which antimony (4.8–6.4%) replaces the tin. These alloys also have a very high silver, nickel, and arsenic content, which strongly suggests that their copper was obtained by smelting a fahlerz copper ore. These ores occur in areas of secondary enrichment within the ore body and are situated below the *gossan* layer. The copper and heavy metals in this layer have reached down into the layer beneath, leaving the iron and silica behind. Thus, fahlerz ores are especially concentrated in copper and other metals, except iron. They were commonly used in the Bronze Age as can be seen by

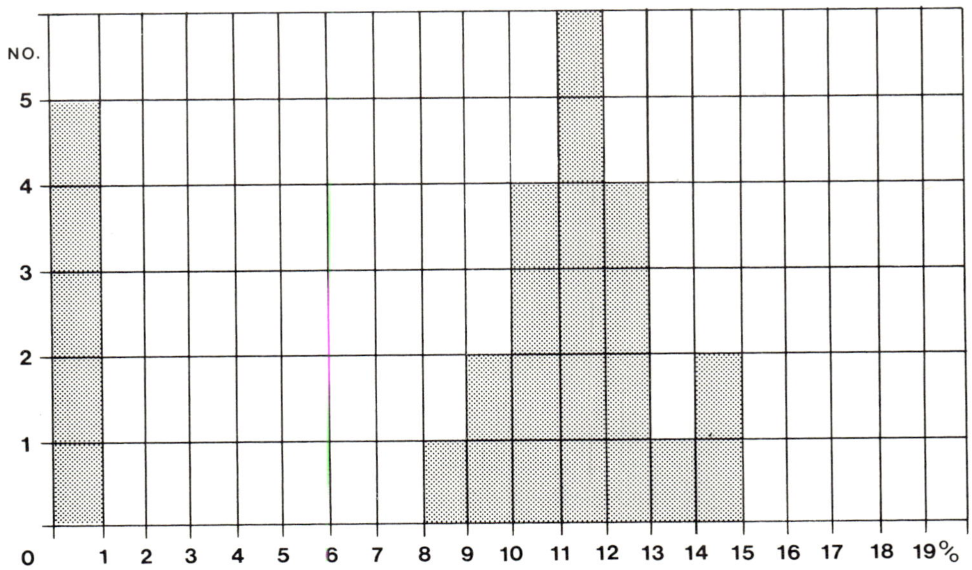

Fig. 1 Tin content of Early Bronze Age bronzes.

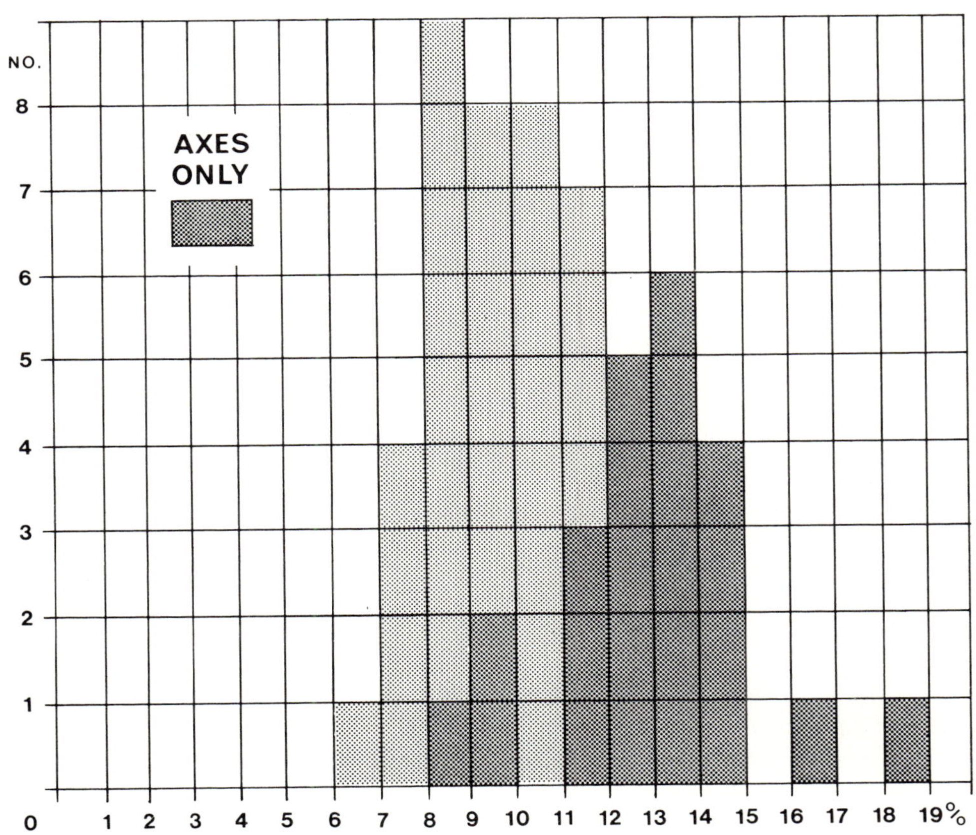

Fig. 2 Tin content of all Middle Bronze Age bronzes.

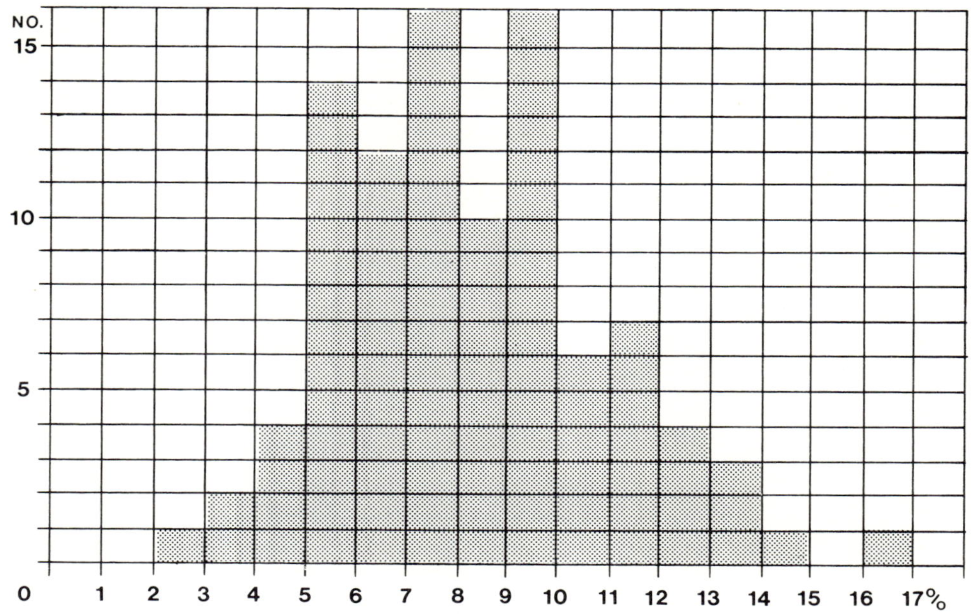

Fig. 3 Tin content of Late Bronze Age bronzes.

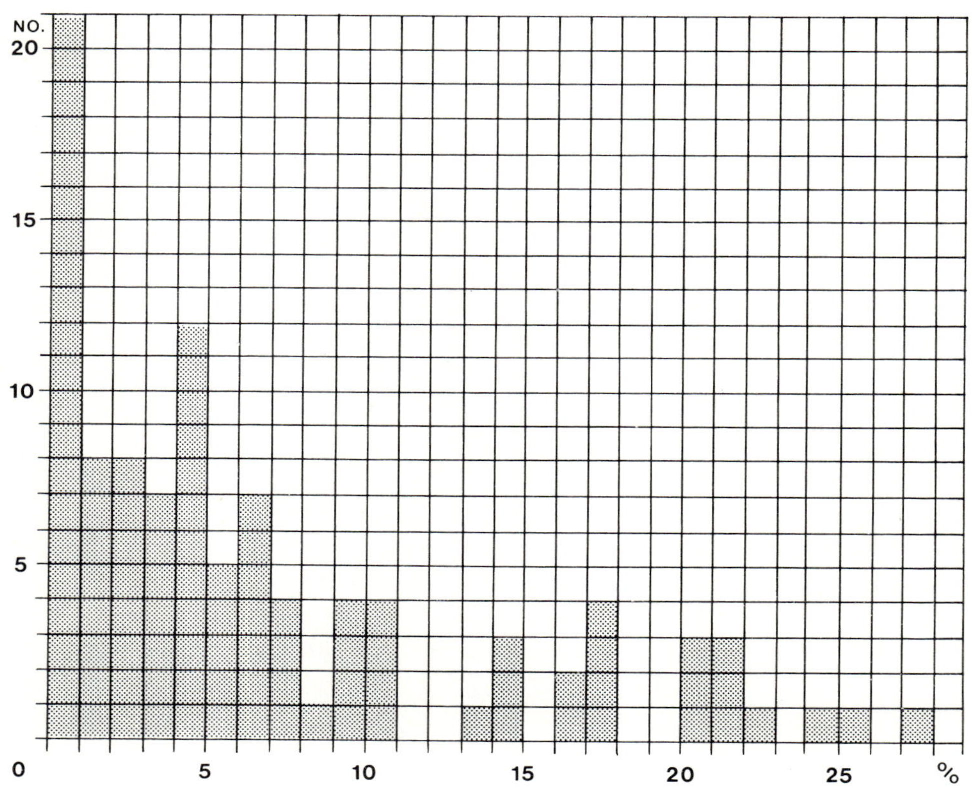

Fig. 4 Lead content of Late Bronze Age bronzes.

inspection of the analytical tables. However, the particular interest of these three artifacts is their exceptionally high antimony content and correspondingly low tin content. This absence of tin in what should otherwise have been a tin bronze shows that the smith was aware that the antimony-rich copper already provided a satisfactory alloy and needed no further additions. Although it is impossible to be sure whether the ore was deliberately chosen to produce such an antimony bronze, or whether the fortuitous use of such an ore produced a metal that was recognized as a bronze before further alloying took place, there is no doubt that these are antimony bronzes. These are the first antimony bronzes to be recognized from Britain, although a few are known from Europe (Craddock 1976), and Bronze Age artifacts of antimony alone have been found at Monte Bradoni in Italy (Trump 1966). Antimony bronze is quite usable, but never popular, probably because it is rather brittle compared with tin bronze.

Three axes (nos. 42, 52, and 72) all contain zinc (from 3.4 to 18.2%). Zinc in bronze alloys is most unusual before the Roman period in Great Britain and strongly suggests that they are not ancient. The unlooped palstave, no. 42, is described in the catalogue as being of "most unusual form".

After the Roman period most alloys used in Britain do contain some zinc, and its absence from the two pins (nos. 183 and 184) suggests they may well be ancient. Similarly the bracelet, no. 178, stylistically of uncertain date, can be placed in the Late Bronze Age or Iron Age on the basis of its composition—leaded bronze being very unusual before this period.

Multiple samples were taken along the length of an axe, a dirk, a rapier, and five swords. Since the weapons were all complete long castings they provided useful samples for examining the degree of segregation that took place in the alloy during casting. The axe had a high lead content. Palstaves were normally cast from the butt and many of the unfinished Breton and Iberian examples still have the filling of the sprue cup, through which the metal was poured, in place on the butt. This casting position is reflected in the ROM example (no. 53) by the gravity segregation of the lead and the normal segregation of the tin. The lead content is higher at the blade end, showing that the lead had time to sink through the mould. The tin content is high at the butt end where the metal, being the last to set, has a higher concentration of the lower melting point component, that is, tin in a tin bronze. Although lead does not enter into solution with the bronze and is therefore the last metal to solidify, its concentration is lower at the butt end because gravity segregation is usually more pronounced than normal segregation, especially if the mould is heated.

The rapier (no. 137), which is unleaded, shows normal segregation of the tin, whereas the dirk (no. 139), also unleaded, has a virtually uniform composition showing that it set very quickly. Because the dirk is a very light casting relative to other objects, this composition would be expected.

Of the five swords (nos. 144–147 and 150), all of which are leaded, two (nos. 144 and 145) show gravity segregation of the lead and normal segregation of the tin; the direction of the concentration gradients shows that they were cast from the butt, in the normal way. Sword no. 150 shows a similar segregation of tin and lead, but the concentration gradient directions are reversed, suggesting that it may have been cast from the tip of the blade—a method of casting for which there is some evidence.

Sword no. 147 shows normal segregation of the tin but, suprisingly, no segregation of the lead. Coghlan (1971) investigated the variation in the lead content along the blade of a leaf-shaped sword with about 4.2 per cent lead and also found no segregation; although Moss (1953), in a similar investigation of a sword with about 12 per cent lead, found considerable gravity segregation. Coghlan concluded that the sword with no segregation may have been cast lying on its side, but this explanation does not account for the normal segregation of tin found in the present instance. Sword no. 146 shows some segregation of the lead, but not of the tin, suggesting that cooling in the mould was relatively fast.

The rivets associated with the rapiers, swords, and the shield (no. 163) were analysed where extant. Generally they have a broadly similar composition to the metal they are attached to, suggesting that no special composition was preferred.

Sample Locations on the Shield (no. 163)

Eight samples were taken. Their locations, given below, are described using the orientation of the rear view, as illustrated above (p. 21).

A Handle, at extreme lower left-hand corner.
B Lower handle rivet.
C Upper handle rivet.
D Centre of upper left-hand tab.
E Upper left-hand tab rivet.
F Centre of upper right-hand tab.
G Upper right-hand tab rivet.
H Outside edge, from the rear, at four o'clock.

Note: the two lower tabs and rivets are undoubtedly modern, the rivets being lathe-turned.

Table listing results of the analyses (percentages of the elements encountered)

Flat and Flanged Axes

ROM	Cu	Pb	Sn	Ag	Fe	Sb	Ni	Au	Co	As	Bi	Zn
† 1	99.0	.050		.005	.019	.040	.025			.050	.050	
2	89.0	.520	10.2	.180	.037	.350	.008		.004	.350	.020	
3	96.5	.040		.290	.020	.510	.015		.002	1.75	.005	
4	96.5	.200	.060	.230	.010	.700	.010			1.20	.005	
5	87.0	1.80	11.9	.090	.010	.100	.015			.130	.010	
6	88.0	.030	10.7	.045	.019	.030	.012		.010	.085	.010	
7	88.5	.030	11.3	.030	.020	.020	.007		.003	.350	.007	
8	89.0	.003	11.1	.002	.020	.100	.005			.030	.010	
9	89.5	.009	9.10	.090	.040	.025	.010			.300	.020	.004
10	87.0	.003	12.8	.015	.011	.030	.002		.002	.020	.010	
11	87.5	.025	12.4	.080	.020	.035	.004			.450	.010	.002
* 12	99.0	.010		.430	.020	.150	.005			.530		
13	88.5	.010	11.3	.004	.025	.025	.007		.006	.050	.007	
* 14	87.0	.007	12.5	.005	.040	.015	.002			.010	.020	
15	87.5	.030	11.8	.090	.026	.030	.008		.002	.300	.010	
16	89.5	.010	11.0	.110	.022	.150	.003		.006	.110	.007	
17	91.0	.030	8.60	.023	.040	.035	.010			.280	.007	
18	85.5	.050	14.3	.005	.050	.040	.220		.003	.020	.020	
*19	87.0	.025	13.4	.011	.014	.015	.005			.010	.015	
20	86.5	.045	12.9	.008	.020	.020	.002			.012	.015	
21	85.5	.190	14.7	.001	.014	.040	.050			.005	.010	
22	85.5	1.10	10.8	.075	.012	.370	.120		.002	.750	.015	.002
23	89.0	.070	9.80	.100	.011	.080	.003			.050	.005	

Palstaves and Winged Axes

ROM	Cu	Pb	Sn	Ag	Fe	Sb	Ni	Au	Co	As	Bi	Zn
24	83.5	.100	16.0	.019	.024	.100	.520		.030	.230	.010	.003
25	84.0	1.90	13.2	.050	.018	.015	.350		.016	.750	.005	
26	73.0	17.5	9.00	.110	.008	.220	.075		.010	.200	.010	
27	81.5	.260	18.2	.030	.070	.220	.440		.030	.260	.015	.012
28	85.0	.090	14.5	.008	.020	.040	.040		.010	.240	.005	
29	84.5	.170	13.9	.024	.160	.130	.530		.032	.200	.005	.010
30	88.5	.160	11.0	.032	.010	.050	.130		.010	.220	.010	.010
31	85.0	1.25	12.1	.045	.016	.035	.250		.017	.620	.005	
32	84.0	.410	14.8	.037	.415	.060	.260		.041	.750	.005	.004
33	88.0	.300	11.1	.045	.055	.030	.250		.013	.700	.005	
34	89.0	.690	9.40	.038	.020	.060	.210		.018	.600	.007	
35	84.5	.090	13.4	.018	.020	.110	.410		.025	.270	.005	
36	90.0	.060	8.40	.039	.190	.050	.190		.020	.570	.007	
37	84.5	2.35	12.0	.020	.019	.030	.220		.015	.550	.005	.030
38	86.0	.600	13.0	.030	.055	.045	.190		.015	.480	.005	
39	88.5	.420	10.4	.040	.018	.045	.210		.017	.550	.013	
40	86.0	.009	14.1	.032	.008	.020	.005			.020	.005	

Continued on page 72.

Palstaves and Winged Axes (cont.)

ROM	Cu	Pb	Sn	Ag	Fe	Sb	Ni	Au	Co	As	Bi	Zn
41	87.5	.120	11.0	.050	.018	.120	.260		.015	.200	.010	
† 42	82.5	8.20	5.50	.080	.190	.100	.035			.240	.160	3.40
43	89.0	.850	9.60	.040	.022	.070	.185		.015	.550	.007	
44	85.5	.780	12.0	.040	.025	.030	.270		.030	.500	.007	
45	86.5	.110	12.2	.034	.010	.030	.075		.009	.210	.010	.005
46	88.0	.060	12.1	.013	.065	.017	.075		.007	.300	.001	
47	86.5	.015	13.6	.016	.016	.030	.030			.060	.015	
48	86.5	.035	14.1	.026	.020	.020	.006			.013	.005	
49	88.0	.170	11.1	.040	.090	.030	.210		.030	.390	.005	.003
50	86.0	.180	13.2	.030	.045	.019	.190		.014	.470	.010	
51	86.0	.150	12.6	.045	.010	.035	.250		.020	.310	.003	
† 52	82.0	6.30	5.20	.095	.400	.050	.040			.230	.005	5.10

Socketed Axes

ROM	Cu	Pb	Sn	Ag	Fe	Sb	Ni	Au	Co	As	Bi	Zn
53A	70.0	25.7	5.20	.070	.004	.035	.050		.007	.120	.015	
53B	69.0	25.9	4.90	.070	.003	.100	.050		.006	.100	.007	
54	87.5	5.80	5.20	.085	.010	.130	.075		.008	.220	.020	
55	85.5	6.10	8.10	.075	.010	.140	.050		.003	.080	.010	
56	86.5	1.30	11.8	.065	.007	.090	.050			.030	.005	
57	82.0	10.0	6.50	.210	.002	.400	.100		.002	.130	.003	
58	72.0	22.3	5.40	.075	.030	.020	.045		.002	.050	.020	
59	85.5	2.90	9.20	.057	.015	.360	.045		.030	.800	.050	
* 60	91.5	2.10	5.40	.100	.010	.180	.070			.070	.005	
61	79.5	9.80	10.6	.160	.007	.300	.100		.002	.065	.010	
62	88.5	.110		2.65	.010	6.40	.650	.005	.003	.950	.075	
63	83.0	.390	16.4	.023	.008	.020	.010			.030	.007	
64	83.5	1.40	14.3	.070	.010	.060	.070		.010	.020	.002	
65	88.0	3.10	8.50	.080	.005	.140	.085		.003	.020	.010	
66	85.5	6.20	9.20	.065	.004	.150	.060		.002	.035	.005	
67	87.0	3.90	8.50	.140	.009	.270	.095		.003	.045	.010	
68	85.5	2.90	11.4	.300	.010	.460	.150		.005	.170	.005	
69	84.0	3.90	11.6	.100	.004	.140	.060		.001	.100	.007	
70	86.5	4.70	6.90	.250	.006	.430	.140		.005	.120	.005	
71	90.0	.100	9.40	.040	.030	.060	.100	.020	.035	.180	.005	
† 72	70.5	7.20	1.90	.095	1.40	.135	.160		.005	.150	.090	18.2
73	90.0	1.30	8.00	.060	.013	.110	.040		.003	.040	.005	
74	86.0	4.60	9.20	.085	.030	.130	.060		.003	.140	.005	
75	86.5	6.20	6.80	.150	.008	.230	.070		.002	.080	.010	
76	87.5	1.00	12.4	.040	.005	.090	.030		.003	.090	.003	
77	85.0	4.10	9.30	.080	.010	.140	.055			.150	.010	
78	89.0	3.50	7.10	.080	.006	.090	.055	.004	.002	.050	.007	
79	83.0	3.15	13.9	.140	.008	.150	.090		.004	.050	.015	
80	84.0	5.10	9.50	.100	.035	.170	.090		.003	.030	.005	

Spearheads

ROM	Cu	Pb	Sn	Ag	Fe	Sb	Ni	Au	Co	As	Bi	Zn
81	89.0	.150	9.80	.012	.050	.050	.032		.006	.120	.005	
82	87.0	.410	11.8	.028	.020	.040	.180		.014	.450	.005	
83	91.0	.070	8.00	.040	.240	.060	.150		.020	.750	.005	
84	86.5	.060	11.4	.011	.250	.180	1.20		.040	.220	.005	
85	88.0	1.90	9.50	.085	.009	.200	.080		.005	.060	.010	
86	90.0	.060	10.6	.025	.009	.060	.080		.020	.160	.010	
87	84.5	9.30	4.80	.170	.100	.430	.100		.010	.350	.005	
88	87.5	.010	11.5	.015	.030	.040	.230		.006	.210	.005	
89	74.0	20.2	5.60	.075	.007	.220	.050		.005	.160	.003	
90	72.5	21.1	6.10	.040	.028	.110	.027		.005	.045	.003	
91	91.0	2.50	6.10	.160	.017	.500	.130		.017	.170	.005	
92	84.0	7.30	7.40	.130	.015	.410	.100		.015	.160	.003	
93	71.5	21.4	5.80	.180	.040	.570	.720		.018	.250	.005	
94	71.5	21.2	7.20	.090	.023	.300	.070		.007	.030	.005	
95	79.0	16.4	3.60	.095	.010	.300	.085		.003	.240	.005	
96	71.5	20.0	7.80	.065	.040	.150	.065		.020	.080	.005	
97	75.0	17.8	5.60	.110	.010	.200	.200		.060	.250	.010	
98	90.0	2.00	6.60	.180	.040	.600	.130		.015	.350	.010	
99	90.0	1.80	6.30	.170	.050	.550	.140		.015	.250	.010	
* 100	81.5	13.1	4.60	.025	.080	.270	.060		.008	.030	.005	
* 101	80.0	14.1	4.30	.090	.020	.280	.055		.005	.300	.003	
102	67.0	24.5	7.80	.012	.090	.020	.001		.002	.015	.003	
103	72.5	20.0	6.10	.105	.015	.330	.110		.020	.150	.005	
104	77.5	14.6	6.90	.140	.025	.450	.070		.008	.200	.005	
105	87.5	6.30	5.10	.190	.050	.470	.120		.015	.250	.005	
106	76.5	17.3	5.80	.140	.007	.430	.095		.012	.290	.005	
107	81.5	9.80	7.50	.015	.010	.130	.100		.005	.050	.005	
108	89.5	.420	9.10	.025	.020	.065	.020		.002	.030	.005	.003
109	89.0	1.60	6.50	.370	.025	1.65	.690		.180	.700	.010	
110	86.0	5.50	8.50	.080	.500	.110	.050			.080	.010	
111	88.0	.015	11.3	.020	.030	.012	.400		.023	.130	.010	
112	88.0	.160	11.2	.050	.070	.140	.180	.010	.030	.130	.010	.006
113	90.0	.090	9.30	.014	.071	.070	.520		.040	.280	.005	
114	90.0	.600	8.70	.035	.050	.110	.200		.025	.250	.005	
115	90.0	.060	9.00	.040	.013	.050	.040		.005	.150	.015	
116	91.0	.030	7.90	.310	.025	.170	.007			.040	.008	
117	88.0	1.00	9.60	.040	.015	.030	.220		.020	.260	.005	
118	89.5	.085	10.2	.029	.110	.070	.150		.025	.750	.007	
119	92.0	.070	7.10	.060	.035	.060	.170			1.40	.015	
120	92.0	.005	7.80	.020	.020	.080	.070		.020	.800	.005	
121	90.0	.120	9.50	.190	.010	.075	.120		.010	.170	.005	
122	80.0	9.30	9.80	.085	.006	.250	.070		.008	.120	.003	
* 123	90.0	.140	8.60	.045	.010	.140	.120		.025	.160	.005	

Continued on page 74.

Tanged Knives

ROM	Cu	Pb	Sn	Ag	Fe	Sb	Ni	Au	Co	As	Bi	Zn
124	88.0	.100	10.0	.050	.025	.200	.290		.020	.050	.007	
125	91.5	1.20	8.10	.060	.020	.080	.040		.010	.220	.005	
126	91.0	.070	8.00	.040	.035	.035	.110		.010	1.30	.007	
127	88.5	.065	10.8	.040	.012	.140	.100		.010	.100	.005	
128	90.5	.020	8.90	.020	.050	.055	.005			.060	.005	
129	89.5	.250	8.50	.040	.020	.220	.180		.007	.370	.007	
130	89.5	.180	9.00	.035	.040	.085	.200		.015	.600	.005	
131	92.0	.120	7.40	.034	.025	.050	.160		.010	.450	.005	

Socketed Knives

ROM	Cu	Pb	Sn	Ag	Fe	Sb	Ni	Au	Co	As	Bi	Zn
132	88.5	2.70	7.90	.085	.012	.070	.090		.030	.170	.015	
133	89.5	4.70	5.50	.090	.006	.130	.060		.002	.030	.003	
134	87.0	4.40	7.10	.090	.007	.150	.060			.080	.005	

Dirks and Rapiers

ROM	Cu	Pb	Sn	Ag	Fe	Sb	Ni	Au	Co	As	Bi	Zn
135A	88.0	.080	10.6	.025	.015	.035	.120		.008	.390	.005	
135B	85.5	.750	12.8	.045	.030	.015	.230		.025	.650	.005	
136	89.0	.110	10.0	.025	.040	.070	.230		.020	.200	.005	
137A	88.0	.085	11.9	.020	.025	.070	.430		.025	.300	.005	
137B	87.0	.100	11.7	.025	.030	.090	.440		.020	.350	.003	
138	89.5	.150	10.0	.050	.028	.100	.120		.085	.400	.005	
139A	90.5	.090	8.00	.055	.020	.080	.095		.010	.150	.002	
139B	91.0	.060	8.00	.035	.030	.100	.090		.020	.250	.005	
140	87.5	1.40	10.8	.030	.014	.070	.030		.003	.020	.005	
141	91.0	1.00	7.00	.060	.018	.150	.015			.020	.005	

Swords

ROM	Cu	Pb	Sn	Ag	Fe	Sb	Ni	Au	Co	As	Bi	Zn
142	80.0	6.80	12.2	.110	.014	.210	.065		.002	.130	.005	
143	83.0	7.80	7.80	.150	.015	.500	.125		.016	.320	.001	
144A	86.5	4.35	9.00	.120	.019	.140	.090		.002	.150	.005	
144B	87.0	4.50	7.60	.090	.014	.130	.080			.200	.005	
144C	86.0	5.30	8.70	.105	.017	.130	.085			.160	.010	
145A	86.5	4.60	8.70	.160	.012	.220	.080		.005	.100	.005	
145B	86.0	6.50	7.70	.110	.011	.270	.085		.006	.120	.005	
145C	86.0	6.00	7.30	.120	.002	.090	.085		.015	.120	.007	
145D	86.0	5.40	8.60	.120	.050	.200	.085		.005	.070	.005	
145E	84.0	5.70	9.40	.130	.010	.240	.085		.005	.130	.005	
146A	83.5	6.30	7.90	.160	.025	.650	.210		.070	.120	.005	
146B	83.0	7.40	7.90	.160	.028	.650	.210		.065	.280	.008	
146C	83.5	7.20	8.00	.170	.028	.650	.220		.070	.310	.010	

147A	73.0	17.0	9.50	.145	.060	.540	.115	.005	.017	.210	.002	
147B	75.0	17.6	6.20	.145	.040	.550	.120		.015	.370	.005	
147C	75.0	17.0	6.40	.150	.015	.550	.130		.020	.250	.005	
147D	77.0	15.4	6.30	.190	.007	.620	.160		.025	.300	.005	
147E	64.5	27.8	7.60	.070	.025	.170	.050		.004	.080	.003	
148	90.5	1.00	7.50	.110	.020	.035	.015			.050	.010	
149A	85.5	.670	13.6	.210	.035	.110	.380		.025	.240	.020	.003
149B	88.5	.800	9.20	.075	.007	.110	.090		.025	.220	.010	
149C	88.0	.750	10.7	.070	.010	.190	.100	.005	.025	.210	.005	
150A	88.0	4.50	7.10	.085	.005	.160	.070		.002	.130	.005	
150B	89.0	4.30	7.10	.095	.008	.050	.065			.220	.005	
150C	86.0	4.20	8.60	.090	.020	.050	.070			.200	.010	
151	89.5	.320	9.00	.045	.010	.030	.030			.027	.007	
152	87.5	3.10	8.60	.050	.008	.025	.150		.013	.100	.005	
153A	87.5	3.50	8.10	.080	.006	.100	.050		.002	.090	.005	
153B	90.0	2.40	6.90	.100	.015	.120	.065			.070	.010	
153C	89.5	2.50	6.70	.110	.008	.120	.060			.045	.010	
153D	89.5	2.50	6.60	.120	.011	.100	.060			.110	.010	

Miscellaneous

ROM	Cu	Pb	Sn	Ag	Fe	Sb	Ni	Au	Co	As	Bi	Zn
154	89.5	.015	10.4	.008	.040	.080	.005		.003	.100	.010	
155	84.5	4.10	2.30	1.30	.004	4.80	1.06		.035	1.60	.010	
156	84.0	5.80	10.2	.040	.006	.110	.020			.200	.005	
157	89.0	4.90	4.50	.350	.520	1.00	.150			.300	.050	
158	94.0	.270	3.80	.045	.470	.120	.010			1.30	.007	.025
159	89.5	.700	6.80	.190	.015	.850	.210		.070	.370	.015	.005
160A	90.0	.190	8.80	.090	.040	.350	.030			.100	.005	.013
160B	90.5	.220	8.50	.090	.030	.380	.070			.180	.005	.013
160C	90.5	.240	8.60	.120	.200	.300	.025			.070	.007	.110
160D	90.0	.330	9.00	.080	.220	.270	.025			.150	.007	.200
160E	89.0	6.90	4.10	.090	.018	.330	.030	.040		.210	.007	.025
*161	90.0	.180	1.50	.800	.280	5.30	.070		.030	2.50	.055	.005
162	86.0	2.00	10.6	.065	.110	.230	.050			1.10	.010	
163A	87.0	.060	12.4	.065	.017	.110	.010		.015	.300	.007	
163B	84.5	2.00	12.4	.037	.010	.010	.290		.045	.280	.005	.004
163C	89.0	.350	10.1	.020	.010	.080	.150		.020	.430	.007	
163D	85.0		13.6	.050	.045	.020	.045			.750	.010	
163E	87.0	.015	12.5	.015	.040		.040			.300	.005	
163F	86.0		13.4	.010	.015		.022		.030	.500	.010	
163G	87.0	.007	12.8	.045	.040		.035		.010	.100	.005	.010
163H	87.0	.015	12.0	.015	.070		.020			.100		
164	85.0	5.00	9.70	.070	.010	.150	.010		.003	.070	.005	
165	84.0	4.30	10.8	.040	.005	.020	.380			.150	.007	
166	88.5	2.30	9.00	.085	.030	.350	.045			.400	.010	.010
167	89.5	4.00	6.80	.070	.010	.070	.075		.023	.220	.030	
168	98.0	.035		.210	.008	.400	.008			1.35	.010	

Continued on page 76.

Hoards

ROM	Cu	Pb	Sn	Ag	Fe	Sb	Ni	Au	Co	As	Bi	Zn
169	83.0	10.9	6.20	.135	.006	.230	.090			.130	.003	
170	84.0	7.60	7.20	.130	.015	.200	.080		.002	.110	.015	
171	82.0	4.10	13.4	.055	.010	.090	.055		.002	.120	.010	
172	85.5	5.00	9.20	.070	.010	.040	.050		.003	.080	.010	
173	86.5	2.45	10.2	.028	.015	.040	.025		.001	.040	.005	
174A	83.5	4.20	10.6	.220	.030	.400	.150	.009		.330	.007	
†174B	8.00	.080		.200	.400	2.60	.055	88.0	.015		.045	.030
175	80.0	10.0	8.60	.150	.018	.300	.090		.003	.200	.010	
176	80.0	10.9	7.90	.150	.013	.330	.100			.500	.005	

Ornaments

ROM	Cu	Pb	Sn	Ag	Fe	Sb	Ni	Au	Co	As	Bi	Zn
177	88.0	.045	11.1	.015	.070	.085	.370		.030	.420	.007	
178	81.0	11.0	6.70	.160	.050	.360	.140	.100	.025	.200	.010	
181	78.0	16.8	5.50	.060	.003	.100	.080	.010	.003	.200	.005	
182	76.5	14.4	8.30	.080	.070	.200	.050			.300	.007	
183	94.0	.170	5.00	.400	.035	.100	.210		.020	.150	.005	.020
184	96.0	.240	2.90	.055	.130	.110	.170		.030	.080	.010	.045

* Mn detected, 0.001% in nos. 14, 19, and 60; 0.002% in nos. 12, 100, 101, and 123; 0.003% in no. 161.

† Cd detected, 0.001% in nos. 1, 42, and 72; 0.002% in no. 52; 0.025% in no. 174B

Provenance Index

Each location is followed by the catalogue number(s) of bronzes found there.

Britain

Ireland

Bibliography

ANNABLE, F.K. and D.D.A. SIMPSON

1964 *Guide catalogue of the Neolithic and Bronze Age collections in Devizes Museum*. Wiltshire Archaeological and Natural History Society, Devizes.

ASHBEE, P.

1960 *The Bronze Age round barrow in Britain*. Phoenix, London.

1970 *The earthen long barrow in Britain*. Dent, London.

BAILEY, A.R.

1964 *A text book of metallurgy*, 389–391. Macmillan, London.

BINCHY, E.

1967 "Irish razors and razor-knives of the Middle Bronze Age", in Rynne, E., ed., *North Munster Studies*, 43–60. Thomond Archaeological Society, Limerick.

BRIGGS, C.S.

1976a "The indigenous minerals and metallurgy of the earliest Irish Bronze Age", *Irish Archaeological Research Forum*, 3, 9–15.

1976b "Notes on the distribution of some raw materials in later prehistoric Britain", in Burgess and Miket (1976), 267–282.

BRITTON, D.

1963 "Traditions of metal-working in the Later Neolithic and Early Bronze Age of Britain: Part 1", *Proceedings of the Prehistoric Society*, 29, 258–325.

BROWN, M.A. and A.E. BLIN-STOYLE

1959 "A sample analysis of British Middle and Late Bronze Age material, using optical spectrometry", *Proceedings of the Prehistoric Society*, 25, 188–208.

BURGESS, C.B.

1968a *Bronze Age metalwork in northern England, c. 1000 to 700 B.C.* Oriel Press, Newcastle-upon-Tyne.

1968b "Bronze Age dirks and rapiers as illustrated by examples from Durham and Northumberland", *Transactions of the Architectural and Archaeological Society of Durham and Northumberland*, n.s. 1, 3–26.

1969a "The Later Bronze Age in the British Isles and north-western France", *Archaeological Journal*, 125, 1–45.

1969b "Some decorated socketed axes in Canon Greenwell's Collection", in "Studies in Commemoration of William Greenwell 1820–1918", *Yorkshire Archaeological Journal*, 42, 267–272.

1970 "Breton palstaves from the British Isles", *Archaeological Journal*, 126, 149–153.

1974 "The Bronze Age", in Renfrew (1974), 165–232.

BURGESS, C.B. and J.D. COWEN

1972 "The Ebnal hoard and Early Bronze Age metal-working traditions", in Lynch, F. and C.B. Burgess, eds., *Prehistoric man in Wales and the West*. Adams and Dart, Bath.

BURGESS, C.B. and R. MIKET, EDS.

1976 *Settlement and economy in the third and second millennia B.C.* British Archaeological Reports, Oxford.

BUTLER, J.J. and I.F. SMITH

1956 "Razors, urns and the British Middle Bronze Age", *University of London Institute of Archaeology Annual Report*, 12, 20–52.

CHASE, W.T.

1974 "Comparative analysis of archaeological bronzes", in Beck, C.W., ed., *Archaeological chemistry*. American Chemical Society, Washington.

CLARKE, D.L.

1970 *Beaker pottery of Great Britain and Ireland*, 2 vols. Cambridge University Press.

COGHLAN, H.H.

1971 A metallurgical examination of a bronze sword found in the Thames at Battersea. *Bulletin of the Historical Metallurgy Group*, 5, 71–74.

COGHLAN, H.H. and H. CASE

1957 "Early metallurgy of copper in Ireland and Britain", *Proceedings of the Prehistoric Society*, 23, 91–123.

COLES, J.M.

1959–60 "Scottish Late Bronze Age metalwork", *Proceedings of the Society of Antiquaries of Scotland*, 93, 16–134.

1962 "European Bronze Age shields", *Proceedings of the Prehistoric Society*, 28, 156–190.

1963–64 "Scottish Middle Bronze Age metalwork", *Proceedings of the Society of Antiquaries of Scotland*, 97, 82–156.

1964 "A Late Bronze Age find from Pyotdykes, Angus, Scotland, with associated gold, cloth, leather and wood remains", *Proceedings of the Prehistoric Society*, 30, 186–198.

1966 Review of Langmaid (1966), in *Proceedings of the Prehistoric Society*, 32, 369.

1968–69 "Scottish Early Bronze Age metalwork", *Proceedings of the Society of Antiquaries of Scotland*, 101, 1–110.

1971 "Bronze Age spearheads with gold decoration", *Antiquaries Journal*, 51, 94–95.

COLES, J.M. and J. TAYLOR

1971 "The Wessex Culture: a minimal view", *Antiquity*, 45, 6–14.

COWEN, J.D
1951 "The earliest bronze swords in Britain, and their origins on the Continent of Europe", *Proceedings of the Prehistoric Society*, 17, 195–213.
1967 "The Hallstatt sword of bronze: on the Continent and in Britain", *Proceedings of the Prehistoric Society*, 33, 377–454.

CRADDOCK, P.T.
1976 "The composition of copper alloys used in the classical world. Part 1. The Greek Bronze Age", *Journal of Archaeological Science*, 3, 93–113.
1977 "The composition of Iberian Bronze Age metalwork in the British Museum", in Oddy, W.A., ed., *Aspects of early metallurgy*. Historical Metallurgy Society and the British Museum Research Laboratory, London.

CUNLIFFE, B
1974 *Iron Age communities in Britain*. Routledge, London.

CURRELLY, C.T.
1956 *I brought the ages home*, reprinted 1976. Royal Ontario Museum, Toronto.

DAVEY, P.J.
1973 "Bronze Age metalwork from Lincolnshire", *Archaeologia*, 104, 51–127.

EOGAN, G.
1964 "The Later Bronze Age of Ireland in the light of recent research", *Proceedings of the Prehistoric Society*, 30, 268–351.
1965 *Catalogue of Irish bronze swords*. Dublin.

EOGAN, G. and M. HERITY
1976 *Ireland in prehistory*. Routledge, London.

EVANS, J.
1881 *The ancient bronze implements, weapons, and ornaments, of Great Britain and Ireland*. Longmans, Green, London.

FOX, C.
1923 *The archaeology of the Cambridge region*. Cambridge University Press.
1939 "The socketed bronze sickles of the British Isles", *Proceedings of the Prehistoric Society*, 5, 222–248.

GERLOFF, S.
1975 *The Early Bronze Age daggers in Great Britain, Prähistorische Bronzefunde*, Abt. 6, Bd. 2. Munich.

GREENWELL, W. and W.P. BREWIS
1909 "The origin, evolution, and classification of the bronze spearhead in Great Britain and Ireland", *Archaeologia*, 61, 439–472.

HARBISON, P.
1969a *The axes of the Early Bronze Age in Ireland, Prähistorische Bronzefunde*, Abt. 9, Bd. 1. Munich.
1969b *The daggers and the halberds of the Early Bronze Age in Ireland, Prähistorische Bronzefunde*, Abt. 6, Bd. 1. Munich.

HARDING, D.W.
1974 *The Iron Age in lowland Britain*. Routledge, London.

HAWKES, C.F.C.
1942 "The Deverel Urn and the Picardy Pin: a phase of Bronze Age settlement in Kent", *Proceedings of the Prehistoric Society*, 8, 26–47.

HUGHES, M.J., M.R. COWELL, and P.T. CRADDOCK
1976 "Atomic absorption techniques in archaeology", *Archaeometry*, 18, 19–38.

LANGMAID, N.
1966 *Bronze Age metalwork in Norwich Castle Museum*. City of Norwich Museums.
1976 *Bronze Age metalwork in England and Wales*. Shire Publications, Princes Risborough, Buckinghamshire.

LANTING, J.N. and VAN DER WAALS, J.D.
1972 "British beakers as seen from the Continent", *Helinium*, 12, 20–46.

LAWRENCE, G.F.
1929 "Antiquities from the Middle Thames", *Archaeological Journal*, 86, 69–98.

MEGAW, B.R.S. and E.M. HARDY
1938 "British decorated axes and their diffusion during the earlier part of the Bronze Age", *Proceedings of the Prehistoric Society*, 4, 272–307.

MOORE, C.N. and M.J. ROWLANDS
1972 *Bronze Age metalwork in Salisbury Museum*. Salisbury Museum.

MOSS, A.A.
1953 *Man*, 53, 4, Table III.

PARKES WEBER, F.
1936 "Nova et Vetera: the Sturge-Kalischer disease, and the Sturge-Weber Panathenaic amphora at Toronto", *British Medical Journal*, April 4, 1936, 708–709.

PIGGOTT, S.
1938 "The Early Bronze Age in Wessex", *Proceedings of the Prehistoric Society*, 4, 52–106.

PRYOR, F.M.M.
1976 "A descriptive catalogue of some ancient Irish metalwork in the collections of the Royal Ontario Museum, Toronto", *Journal of the Royal Society of Antiquaries of Ireland*, 106, 73–91.

RENFREW, C., ED.
1974 *British prehistory: a new outline. Duckworth, London.*

ROWLANDS, M.J.
1976 *The production and distribution of metalwork in the Middle Bronze Age in southern Britain*. British Archaeological Reports, Oxford.

SAVORY, H.N.
1958 "The Late Bronze Age in Wales: some new discoveries and new interpretations", *Archaeologia Cambrensis*, 107, 3–63.

SIMPSON, D.D.A., ED.
1971 *Economy and settlement in Neolithic and Early Bronze Age Britain and Europe.* Leicester University Press.

SMITH, M.A.
1959 "Some Somerset hoards and their place in the Bronze Age of southern Britain", *Proceedings of the Prehistoric Society*, 25, 144–187.

SMITH, R.A.
1919 "A note on the shield from the river Lea", *Proceedings of the Society of Antiquaries of London*, 31, 145–146.

TRUMP, B.V.
1962 "The origin and development of British Middle Bronze Age rapiers", *Proceedings of the Prehistoric Society*, 28, 80–102.
1968 "Fenland rapiers", in Coles, J.M. and D.D.A. Simpson, eds., *Studies in Ancient Europe.* Leicester University Press.

TRUMP, D.
1966 *Central and southern Italy before the Romans.* 72. Thames and Hudson, London.

TYLECOTE, R.F.
1962 *Metallurgy in archaeology.* Arnold, London.
1968 "Metallographic examination of Bronze Age artifacts from the north of England", in Burgess (1968a), 48–56.

WARD-PERKINS, J.B.
1939 "Iron Age metal horses' bits of the British Isles", *Proceedings of the Prehistoric Society*, 5, 173–192.

WILDE, W.R.
1857 *Catalogue of the antiquities. . .in the Museum of the Royal Irish Academy.* Dublin.